THE CHANGING PATTERN OF FARMING
1912 – 2020

Experiences and thoughts of a farmer
in the Chiltern Hills of South Buckinghamshire
during the seven decades of peace in the Western World
following two World Wars

BRYAN K. EDGLEY MBE FRSA

Forewords by
Steve Baker MP FRSA

and by

Katy Dunn,
Editor of *The Clarion*

Published by New Generation Publishing in 2022

First Edition

Paperback ISBN: 978-1-80369-246-3
Hardback ISBN: 978-1-80369-247-0

Book genre: SOC 055000 SOCIAL SCIENCE/Agriculture and Food

This book is set in Times New Roman in 10, 11, 12, 16, and 20 point.

www.newgeneration-publishing.com

New Generation Publishing

Dedicated to the memory of
my dear wife Alison,
with whom I shared every aspect
of our farming and family lives
for 65 years

Contents

Foreword

You have before you an extraordinary book. Extraordinary for its scope and wisdom. Extraordinary for the history it elaborates and the lively, practical style in which that is achieved. Extraordinary because the lives it reflects have been extraordinarily well-lived in the service of other people and the land.

The Changing Pattern of Farming first and foremost covers just that. And what a history it is! Bryan Edgley shares his vivid recollection of the journey from farming by horse and steam to the latest tractors guided and positively steered by GPS. I had never conceived of tractors started by hand with petrol but then running on "tractor vapourising oil". I had not stopped to think of the overheads involved in feeding the horses or the scale of the manual labour required before more sophisticated machinery. That so many workers would be required to support a threshing contractor points to an entirely different rural economy which we cannot recognise today, a poorer one no doubt, but one whose superficial simplicity will attract many.

This book charts the journey of our country from the strip system and common land, through enclosures and the corn laws, and food insecurity during two world wars, right up to our future agricultural policy after leaving the European Union. The Edgleys operated mixed farms of dairy, cereals, pigs and eggs. They integrated adjacent farms and became ever more concentrated on arable farming, bringing in their family as they grew up. This is above all a book about that journey in food production, often expertly described for the lay reader in the pages of *The Clarion*, as you will discover later.

However, as someone in the last generation to grow up without mobile phones, it is a remarkable thing to read Bryan's account of wide-ranging technological change and tumultuous developments in public policy with his wit, insight and good cheer ringing in my ears through every sentence. Bryan's experience of photographic developments could fill another volume. Before antibiotics were discovered, a member of the family perished after an insect bite or minor injury. The range of this book and consequently its importance as a work of history should not be underestimated: for those who come after – to whom technology as yet unimagined will be commonplace – this book will be a treasure trove of relatable history, brought alive by a fine mind, effectively applied throughout a lifetime.

Finally, this is a story of family and of love. It is the story of a beautiful marriage spanning an astonishing 65 years between Bryan and Alison Edgley, bringing up a family and presenting a formidable front in defence of their brood and farm. Whether confronting the redoubtable Barbara Castle MP over the route of the M40 or having practice changed on conveyancing land with chancel repair liabilities, Bryan and Alison were not a couple to trifle with and yet their lives were lived together to the good of all.

After reading this book, I remain an irredeemable free trader but one nevertheless chastened by solid practical lessons in the importance of food security through home growing on farms made profitable by food production, not noble dreams of rewilding. It is an honour to write this foreword to a book of obvious importance today and one which seems likely to grow more significant with age. I commend it to you heartily.

Steve Baker MP FRSA
Wycombe
26th August 2021

Author's Foreword

I hope that this book, *The Changing Pattern of Farming* will be of interest, both now and in the years to come, to those who though not necessarily working in the farming industry like to observe the farming scene. It is not intended to be a technical book for farmers, even though many of my fellow farmers may share my interest in the way that changes are taking place in the methods used to produce food from the land. I have tried to set out some of the many achievements of British farmers, largely in the post-World War II years, together with their concerns and problems, whether technical or political, for the future.

The articles reproduced in Chapter 5 were first published in '*The Clarion*', the magazine published by The Lane End Parish Council. It has a print run of 1,750 copies each quarter and is circulated free of charge to all the houses within the Parishes of Lane End, Cadmore End, Bolter End, Wheeler End and Moor End on the Chiltern Hills, midway between High Wycombe, Marlow and Stokenchurch in Buckinghamshire. The Editor of The Clarion, Katy Dunn, has transformed this parish magazine during the past ten years from being a small quarterly publication on poor quality paper, reporting only on those matters which any Parish Council is obliged to report to its parishioners, into being a parish magazine covering matters of local and general interest printed on good quality paper with full colour graphics and photographs.

I am fortunate in having been able to cover changes in lowland farming in England over the past 100 years, since my late father, Roy Edgley, started recording the farming scene at Oxshott, Surrey in photographs taken on glass quarter-plate negatives when he was a schoolboy, under the expert tuition of my late Great Uncle George Martin – who for a time in the late 1890s had been a professional photographer in New Zealand. My father, although a solicitor by profession, retained his interest in farming matters throughout his life and gave me nearly one hundred of those glass quarter-plate negatives in two wooden cases which I keep in my archive records. Some of these, with captions which I have written from my memory working with such farming systems at Sacrewell, near Peterborough, where I first learned the farming trade from the late William Scott Abbott, are reproduced in the first chapter.

A keen interest in photography does seem to run in my family. I have been taking photographs since 1946 and was awarded the Licentiateship of the Royal Photographic Society in 1977. My oldest son, Paul, who was awarded the Associateship of the Society a few years later, has followed a career in photography teaching the subject within the Art department at Christ College, Brecon from 1996 until his recent retirement in 2020. Paul has assisted me with the preparation of the photographs in the following pages, including making prints of those taken originally by my father over 100 years ago.

Within Chapter 3 I have described our farming here at Kensham Farm from 1955 to 2020. Having been born in 1932, thus being 89 years old now, I may seem on the elderly side for starting to write a book. But I have been encouraged by my nephew Emlyn Coldicott and by my own family to record in book form the changes at Kensham Farm, Cadmore End ever since my dear late wife, Alison, and I started farming its 102 acres here at Michaelmas 1955, six weeks after our wedding at Ashtead, Surrey.

In 2019 Alison and I realised how fortunate we had been to be still together, and still farming the same farm during the 64 years since we married, but that was to end in the following year when, following increasing frailty over the previous couple of years, she died in Summer 2020. Our second son, Charlie, has been running our day-to-day arable cropping of 2,000 acres of combinable crops for several years, and his son, Angus, assists with the practical side of the arable work. My daughter's son, Alex Nelms, assists me in the farm office with management of our diversified farm enterprises and the land agency side of our farms, and solves all the IT problems that our farm office computer may throw up which are beyond the capabilities of Charlie or myself. One of Alex's tasks has been setting up our farm website, which may be seen on www.kenshamfarms.com and describes some of our activities on the overall total of 2,500 acres which we now farm. It has been my great pleasure in the current year to welcome Charlie's son Angus into our Kensham Farms Partnership. I count myself a fortunate man in this modern day and age to be still involved in our farming affairs in partnership with one of my sons and two of my grandsons.

When I learned the farming trade from the late William Abbott at Sacrewell in 1953/54 my first job during the sugar beet harvest was to harness up one of the seven working horses, generally a Suffolk Punch, and put him into the shafts of a one ton tumbril cart, and work with him all day loading the cart by hand and carting the sugar beet to a clamp at the side of the field. These were farming methods similar to those shown in my father's photos taken forty years earlier, now reproduced in Chapter 1.

In the chapters which follow concerning modern farm practice it will be seen how machines, rather than hand labour, carry out most of the work. No hand lifting is necessary on the farm today - if something needs to be lifted, then either hydraulics or electricity, rather than human muscle, will do the lifting.

Our tractors are now equipped with Global Positioning Satellite receivers and screens in the cab, and other farm machines, implements and seed drills have been developed in an increasingly sophisticated way by agricultural engineers. Crop protection products have been developed by agricultural research chemists eliminating much of the former handwork, so that the agricultural work force has changed from a large number of workers who were physically strong to a much-reduced work force of highly skilled tractor drivers and workers. On our Kensham Farms we try to make full use of modern fertilisers and those crop protection products which have so reduced plant diseases that huge increases in farm yields have been achieved, particularly of wheat which is our main crop.

The changes in farming which I have seen, and the daily tasks on the farm, have been through a change which is perhaps as great as the change from sail to steam in shipping. Who knows what the future will bring?

Bryan Edgley
Kensham Farm, Cadmore End July 2021

Chapter 1

Work on a farm at Oxshott, Surrey 1912 – 1916,
captured by early photography on glass quarter plate negatives

My father and grandfather were both solicitors practising as Edwin, Son and Edgley firstly from their office in Trinity Street, Southwark at the end of the 1800s and subsequently in 1924 nearer to the Law Courts in London at Lincoln's Inn, so I am a first-generation farmer. However, my father had been keenly interested in horses and work on farms ever since his school days just before and during World War I. At that time the Edgley family travelled each summer from the family home in Putney to Oxshott in Surrey in their pony trap drawn by the family pony Gyp, a pony that my grandparents kept in livery stables next to the Northumberland Arms on the corner of the Upper Richmond Road and Dyers Lane in Putney, to spend a lengthy summer holiday at Little Heath Cottage, Oxshott.

My grandmother's brother, my Great-Uncle George, had been a professional photographer in the early days of photography in the late 1800s and for a few years he had specialised in the commercial photography of buildings in New Zealand. When Uncle George returned to England to live at his family's home in Putney just before World War One it had been his pleasure to teach the art and practice of photography to my father during his school holidays.

At that time photographs were taken on glass plates with orthochromatic emulsion then developed one by one in a dish of developer in the darkroom with a dim orange light to which orthochromatic emulsion was not sensitive. Each photograph involved firstly loading the individual sensitive glass plates into a double dark slide plate holder designed to clip onto the back of the camera. My father's early photography was with this system using a quarter plate camera with bellows which I still own. It has a Taylor, Taylor and Hobson anastigmatic lens with focal length of 5.7 inches and maximum aperture of f 6.3 which was manually focussed using a marked rack for the bellows on which the lens was mounted. In those days the photographer had to be familiar with the darkroom work of loading the light sensitive glass quarter plates, each of which was 3.25 by 4.25 inches into the double dark slide holder, then setting up the camera and estimating the distance between subject and camera in order to set the lens to the correct focus. The intensity of the light had to be measured before setting an appropriate aperture and shutter speed. A tripod had to be used to hold the camera steady unless the daylight was so bright that a short exposure of 1/50th of a second or less could be used.

Today digital photography with either a digital camera or mobile phone is easy for those with no technical knowledge and gives the photographer the capability of taking hundreds of photos on the same memory card without reloading the device. However, in the early years of photography great skill was necessary for the photograph to be successful, and heavy loads had to be carried if as many as twenty photographs were to be exposed in a day.

So much for the technical aspects of early photography, I will now turn to the achievements of my late father when he was a schoolboy in taking photographs of farming operations in progress at Oxshott. I still hold in my darkroom store nearly 100 of those glass quarter plates negatives which my father exposed and processed in the years 1912 to 1916. I show below digital prints prepared by my son Paul from my father's original glass plate negatives with captions describing the event or farm work in progress: -

The Edgley pony trap drawn by Gyp crossing a ford through the River Wey near Stoke d'Abernon, Surrey, near to the holiday home of Walter and Rose Edgley at Little Heath Cottage, Oxshott, Surrey.

The Edgley children in 1914 all on their pony Gyp posing for a photograph – Doris Edgley holding the reins with Roy Edgley (the author's father) in the centre and Alan Edgley behind him. Alan Edgley preferred horses and outdoor life to office work, so after training as a bank clerk he made the bold decision of quitting the bank and London to take a one-way passage by sea to Australia to work on a sheep station. However, to the lasting sorrow of his mother, the author's grandmother, he suffered an insect bite or minor injury which became infected. That was in the years before the Australian Flying Doctor service had been formed or the development of antibiotics, so that no available treatment could curtail the infection from which he died on the day after his twenty-third birthday.

The farmstead at Oxshott with two of the draft horses returning to their stables after working on the fields. Both horses are still wearing their bridles and head collars, with a nose bag hooked onto the head collar of the leading bay horse held by the author's father, Roy Edgley.

Ploughing with a single furrow plough drawn by a pair of working horses at Oxshott - the last farming to take place on that farm was just after World War II. The area is now purely residential, with many of the houses built in the post-war years.

Cultivating with tined cultivator drawn by a pair of working horses. The lad on the right is the author's father, Roy.

Pause for a photo call in the harvest field. The self-binder is pulled by a pair of horses and has a large ground wheel which bears the weight of the self-binder, and is geared to sprockets which provide power for the cutter, conveying belts and self-knotting mechanism to tie binder twine around the sheaves. This type of self-knotting mechanism had been developed in the 1880s, then patented in the USA in 1892. One of the sheaves of wheat, which has just been tied with twine before being discharged onto the ground, is on the left.

Loading sheaves of wheat in the harvest field onto the four-wheel cart for transport back to the stackyard. The author's grandmother, Rose Edgley, in long summer dress and hat, is loading one of the sheaves of wheat – probably for the photo call, rather than making a real contribution to the work in progress.

In modern times one tractor will pull a trailer with a capacity of 16 tonnes of wheat, driven by a single tractor driver, who will drive alongside the combine harvester to load the trailer on the run. However, in Edwardian times the four-wheel cart had to be loaded by human muscle tossing up the sheaves with a pitchfork to the experienced worker stacking the trailer with great skill - to avoid a carelessly loaded cart from losing parts of its load before arriving at the stackyard. Health and safety regulations hardly existed in those days, so in the photo we can see the skilled farmworker who had stacked the cart with the author's Uncle Alan and his sister Doris and Grandfather Walter Edgley on the top. The photo has slightly faded at the bottom but the worker who had loaded the cart and would be leading the horse back to the stackyard can be seen seated with the author's father in the foreground and his grandmother Rose standing.

The author's Uncle Alan in the driving seat of a single working horse hay rake, to rake up any stalks of wheat which had been missed by the self-binder. In current 21ˢᵗ century farming it would not be cost effective to trouble with the odd wheat plant or stalk that the modern combine harvester had missed.

Threshing time. The belt driving the threshing drum from the flywheel on the steam engine can be seen at the top of the steam engine. The threshing drum would have been towed by the contractor's steam traction engine from the previous farm where it had been threshing and would then have been set up by the contractor, normally using a spirit level, to ensure that the separating sieves within the threshing drum would have been level from one side to the other to secure an even flow of grain. The contractor would have to arrive at the farm on threshing days at least one hour before the other workers in order to light the furnace and get steam up. At the start of the threshing process the sheaves of wheat are fed evenly by hand into the drum which beats the stalks of crop against the concave to separate the grain from the straw. Then the grains and some chaff run over the separating mechanism of the threshing machine with its coarse sieve, designed so that any stones or large impurities would have stayed on top to be separately discharged. There would also have been a sieve with quite small holes, so that all the good grain stayed on top, but any very small grain and weed seeds would fall through, to be bagged off as tailings. The third separating mechanism would have been a fan, also powered by the steam traction engine, to blow out any chaff – bagging that off was the worst job at threshing time, generally allocated to the youngest member of the threshing gang.

The threshing drum at work. It was not uncommon for as many as eleven farm workers to be at work feeding the threshing drum, bagging off the grain in railway sacks which contained 2.25 hundredweight (cwt), or in metric measure 114.3 kilograms - whereas under modern health & safety legislation the maximum size sack that is allowed by law weighs 25 kilograms. The straw was normally baled with a baler also powered by the steam traction engine, using wire tied bales. For such wire ties, one worker had to feed the wire in from one side of the baler, then a less skilled worker had to catch it, line it up over the length of the bale, then poke it back though between bales so that the first worker could then twist it around itself to form a knot. The author can remember tying many such bales with wire ties when he worked in the threshing gang at Sacrewell, William Scott Abbott's well managed farm at Sacrewell, Thornhaugh near Peterborough in 1953/54.

Feeding the sheaves into the drum from the top of the threshing machine. This involved dismantling the corn rick in the stackyard by first removing its thatched roof (the beautiful thatching on the ricks normally had a life of only three or four months, between harvest and threshing time) then tossing the individual sheaves off the rick up to the farm workers who were feeding the drum. Safety guards were minimal, and by today's health and safety standards the job of feeding the drum would be condemned as being too dangerous.

This photograph was taken by my father of school children at the Maypole, watched by an audience of local folk.

The Maypole is another reminder of times gone by when the Parish Church and its Vicar, and rural landowners, were all involved with the State in setting up Church of England schools, this being the beginning of compulsory education in England. The School Sites Act had been enacted in 1841 to encourage rural landowners to give sites for the construction of Church of England schools in these rural areas where previously there had been no education facilities, other than for families sufficiently wealthy to employ nannies or tutors. Some inducement was given to those landowners, in that the Act made provision that if the site should ever cease to be used for the educational purpose for which it had been given the land would 'revert' back to the descendants of the original landowning benefactor.

The author had personal experience of such a reversion, since at the end of the 1960s a new state Primary School had been built in the village of Lane End, on land that had previously been a part of Norman Archer's progressively managed pig farm, as an integral part of a new Council housing estate which more than doubled the population of Lane End. The author was first elected as Churchwarden of Holy Trinity Church, Lane End in 1972, where one of his earliest duties had been to trace the descendants of the original members of the Elwes family who had donated the land on which the now redundant Church of England Lane End School had been built. The reason for this was that on that day when the Church School ceased to be used for teaching the freehold of the school grounds, now with the Head Teacher's School House and all the classrooms standing on it, reverted to the descendants of the original benefactor who had given the site for the school under the provisions of the School Sites Act of 1841.

This final photo from the original quarter plates exposed by the author's father shows the Edgley family's pony and trap with the author's Grandmother Rose in the centre, Aunt Doris on the left and Great-Uncle George, the family's experienced photographer, on the right.

Chapter 2

Farming before and during World War II
with the author's experiences of farming from 1948 to 1955
and thoughts on Food Security

I was born in 1932 and my personal knowledge of farming, and my photographs, started when I was taken to Kensham Farm during school holidays on visits with my father, Roy Edgley, in 1947.

History always has lessons from the past, which so often are relevant to the future. Consideration of agriculture and politics in 2020 is best started with the Agricultural Revolution in the years 1750 to 1850 during which the English population, which had to be fed from the home-grown food supply, had increased from 5.7 million to 16.6 million.

English Farming in the Middle Ages
In the Middle Ages all the villagers were poor. Those that were manorial tenants held individual arable strips on an open field system, with each manorial tenant holding several strips. Each village tried to produce enough food for all those who lived within the village. Some of the strips would have been on the best land and others on poorer land, so that no one villager had a monopoly of all his strips being on the best land. Until the aftermath of the Black Death (1349) those strip holders would have been called villeins, who owed service to the Lord of the Manor. The cattle owned by the villagers were all grazed together on common land, which was the waste land of the Manor.

Enclosure Acts
The many disadvantages of the strip system, which prevented more productive farming practices from being introduced to grow sufficient food for expanding towns, had to change. This was the driver for specific Private Acts of Parliament from 1760 to 1800 which allowed Enclosure, making it possible to consolidate the land into productive larger landholdings. The Lord of the Manor who owned the freehold would have instigated the private Act in question and the Enclosure Commissioners, who would later have been appointed and would become responsible for making decisions over the future pattern of the fields and allocation to the farm tenants. In 1801 a General Enclosure Act made the process of enclosure cheaper and easier, so that by 1850 very few of the original strips remained. Additional farmland had also been formed when the Dutch drained the East Anglian Fens in the seventeenth century, and more reclaimed land came from former woodland.

Imports from Overseas

However, more changes were to come when overseas countries such as the United States of America developed land that had never previously been under cultivation, from which their surplus grain was exported at prices so low as to undercut the cost of grain produced in Britain. At that stage Government enacted the Importation Act 1815 to protect British farming - but that had the drawback of increasing the price of food to the urban population. After fierce debate between those representing the rural areas, and those representing the growing urban population, these 'Corn Laws' were then repealed by the Importation Act 1846.

For several years after 1846 farming continued to yield workable profits for British farmers and landowners, but soon after that wheat grown cheaply on the newly farmed prairies of the USA was imported in increasing quantity into Britain at a cost that was below cost of production of wheat grown on the smaller fields of Britain. This caused the decline of home-grown food production, accentuated when larger ships were developed thus further reducing the cost of freight. British farming reached such a low point by the 1860s that there was great poverty in most rural areas of Britain, and many farms became derelict in that pre-First World War period.

The First World War leading to Shortage of Food in Britain

Farming in Britain continued at a low ebb from the 1860s until the declaration of war in 1914, in which a primary objective of Germany was to win the war by causing starvation in Britain, hoping that this would lead to its capitulation. German U-Boats were soon so successful in sinking British merchant ships in the convoys which had been bringing food to Britain that food shortages became very real. British farmers were therefore urged to recommence food production and started to receive adequate payment for farm produce. The scarcity of food lead to a change of mood about food security, and so Parliament created the Ministry of Food in 1916, and the following year enacted the Corn Production Act 1917 to guarantee minimum prices for wheat and oats up to 1922. It was at the end of 1917 that this new Government department introduced a food rationing scheme.

In 1919 the Board of Agriculture and the Ministry of Food were merged to form the Ministry of Agriculture and Fisheries. This new Ministry of Agriculture and Fisheries took note of the national feeling after the First World War that a man who had fought for his country should be entitled to retire to a smallholding on British land which would provide him with a livelihood from producing food on that smallholding, and this led to various initiatives collectively called 'Homes for Heroes'. New agricultural laws were passed to make this possible by distributing land to ex-servicemen, with County Councils being given compulsory purchasing powers to requisition land which could be let as smallholdings to ex-servicemen. Subsequently those ex-servicemen could buy the land, requesting the County Council to grant them a mortgage to enable them to purchase the freehold. The council could not refuse such a request without the Minister of Agriculture's permission.

The Agriculture Act 1920 foresaw enduring protection for British farmers, linking such protection to the cost of production. This promised stability enabled agriculture to thrive after the war. However, that was to be short lived. With peace came the reopening of worldwide trade, the price of wheat halved, and in 1921 Government swiftly repealed the Corn Production

Act – at the very time when the support which it could have given was most needed. This repeal of the 1920 Act pushed agriculture back into a depression which became known in farming folklore as 'The Great Betrayal'.

The well-meaning 'Homes for Heroes' initiative did not anticipate this difficulty of making a living from the land or the depression that was to follow, in which many of the new smallholdings failed. Land so allocated at Marlow Bottom at the foot of the Chiltern Hills, only a few miles from Kensham Farm, was an example of such smallholdings for ex-servicemen failing. After most of these smallholdings failed, the area was given the nickname 'Tin Town' before the shacks built on the derelict farmland had been redeveloped as private houses.

The reasoning behind 'The Great Betrayal' was that in the 1920s, continuing into the 1930s, the UK Government could not control worldwide depression, and yet it had a responsibility to avoid British folk from starving. It was politic to get the cheapest food possible from somewhere, whether it was from home farms, or from overseas farms with their cheaper land, labour and food production resources. Thus it was overseas imports that caused British farming to decline yet again; arable crop production nearly ceased, and those farmers who survived largely turned to 'dog and stick' farming – a system in which they did not cultivate any crops, but just kept some livestock for beef or lamb production on the grass meadows.

The Second World War – Food Shortages again in Britain

This decline in agricultural production was so severe that in the mid-1930s Britain was importing 55 million tonnes of food a year, with home production dropping to only 12 million tonnes. At the start of 1940, for the second time in less than three decades, food rationing had to be re-introduced as a result of German U-boats sinking large numbers of merchant ships in the shipping convoys. Rationing in Britain did not completely end until July 1954.

During World War II the government tried to encourage people to grow their own food by means of the 'Dig for Victory' campaign, with posters showing a boot on a spade digging ground, and householders were encouraged to keep rabbits and chickens for the table.

Because so many men were being conscripted into the army it became necessary for much of the work on farms to be carried out by women. That was when Government formed the Women's Land Army, known less formally as 'Land Girls', to work on the farms. Folk living in residential areas were encouraged to dig up their gardens, or take on allotments, to grow vegetables to take the place of the imported vegetables and fruit which had been available in the shops before the war. There was one strange campaign under which the Government responded to a temporary wartime oversupply of carrots by suggesting that the RAF's exceptional night flying was due to eating carotene. This ruse worked to the extent that consumption of carrots increased sharply because people thought carrots might help them to see in the blackout. These measures took the pressure off other food supplies at a time when the Prime Minister suggested that, if necessary, food supplies could take priority over supplies for the military when there was a possibility of famine in the occupied territories after the war.

My grandson Richard was asked at his primary school to write a description in 2003 of my own experiences in World War II. Richard's essay is shown below: -

My Family during World War II

When war was declared in 1939 my grandpa was 7 years old, and he remembers listening to the wireless and being told to take shelter under the dining room table.

From 1939-1940 Grandpa and his sister Diana had to go to boarding school on the south coast at West Preston Manor because everyone thought it would be safer. Then the Germans invaded France and the south coast became the unsafe place, so Grandpa went back to Wimbledon. By then there was no petrol for most people except for doctors or very important people. Grandpa was then a day boy at King's College School, Wimbledon and went by bike. There were often air raids at night and the bombing went on for a very long time. Great Grandpa shored up the ceiling of the strongest room in the house next to the kitchen because it had a concrete ceiling, in which he installed bunk beds with 2 rows of bunks on each side. Every night for months all the family slept in the shelter.

After lots of bombing the next day Grandpa and his friends used to go looking for shrapnel, some pieces were 2" long and half an inch thick. More than once they found other houses in their road had been flattened to rubble by heavy bombing.

There was food rationing and they had some chickens in the garden, which provided plenty of eggs for the family. They dug up the tennis court to plant potatoes and looked after the fruit trees carefully and kept all the fruit. Everything was rationed and there were no bananas.

Grandpa said that there was no Television, and everyone listened to the wireless to hear all the latest news from Europe and also from the Government. At night-time there were blackouts on the curtains and the Air Raid wardens came to check the roads to make sure that no lights could be seen. There were no streetlights.

He was evacuated to Woking for 8 weeks where there was no school, so he went fishing.

The author's experiences of farming from 1947 to 1955

While I was at school at Charterhouse near Godalming, Surrey, I found that my recreational interests were not on the playing fields with ball games, but that rifle shooting on the school ranges and in competitions at Bisley Camp, together with Scouting and estate work in the school grounds were of more interest to me. During school holidays visits with my father to Kensham Farm were a highlight.

My father had bought Kensham Farm situated at Cadmore End, near High Wycombe on the Chiltern Hills, as an investment in October 1946. It had been one of six small farms and 17 cottages totalling 686 acres sold by Lord Parmoor of Frieth at a sale on 29th June 1945 of a part of his Cadmore End Estate. Kensham Farm was subject to a full repairing tenancy granted to John Bird, whose sons, Jack and Frank Bird, were still farming it in 1946 as a traditional small mixed farm. The earliest copy of a Tenancy Agreement in my records to John Bird was at Michaelmas 1913, at the extremely low rent of £77 per annum. This rent was still the same when Kensham Farm was sold for £3,300 freehold in the 1945 Parmoor Estate Sale.

Parmoor Estate Sale – June 1945

Folk who have become accustomed to property values in Britain at the present time, 2021, will be truly astonished at the low level of prices for property sold at public auction at that time in June 1945. My neighbour Richard Smith, who is a coal merchant with his yard at Cadmore End, has given me consent to publish the prices at which his grandfather, Richard Bernard Smith, bought the cottage in which Richard now lives, and another detached cottage, and Rackleys Farm which is now an up-market wedding venue. He paid £3,590 freehold for the three properties, as shown on the Form of Agreement witnessed for the Agent to the Parmoor Estate over a stamp of six old pence and by the purchaser's solicitor over a stamp of two old pence. The individual prices paid were £240 freehold for the semi-detached cottage, £750 for the freehold of the detached cottage and £2,600 for Rackleys Farm with its farmhouse, farm buildings and 54 acres of land, equivalent to £66.48 per acre.

At that 1945 sale Kensham Farm, with its 102 acres, farmhouse and farm buildings was bought by Reginald Simmonds, an estate agent at Maidenhead, for £3,300 freehold, who in turn sold it to my father in 1946. The six small farms totalling 606 acres were sold for a total of £20,250 making the average price just over £33 per acre freehold, each farm consisting of its land with a farmhouse and farm buildings. The cheapest 10 of the 17 cottages were sold at an average price of £196 freehold, albeit they were unimproved, mostly without mains water, electricity or interior plumbing system but having an external earth closet. It is interesting to reflect on the reason why the farms and cottages at that 1945 sale were sold for such low prices; those were uncertain times in that it was less than two months since victory in Europe, VE Day, had been announced on 8th May 1945, and World War II continued until the surrender of Japan at VJ Day on 2nd September 1945.

Prior to the Parmoor Estate Sale in 1945 all the farms on the Estate had been let to farming tenants on full-repairing leases at low rents, making the tenant farmers responsible for building repairs which would normally have been the responsibility of the landlord. The system of full repairing leases could be satisfactory if coupled with regular inspection of condition from the landlord's agent to check that the buildings were being correctly maintained.

However, farming between the wars in the 1920s and 1930s had not been prospering and tenant farmers, even those on low rents, were having a struggle to avoid insolvency, leaving no spare funds with which to buy building materials, and certainly not enough money to employ a builder for carrying out repairs. This resulted in the gradual deterioration of the condition of farmhouses and farmsteads in many parts of Britain, including those on the Parmoor Estate. This can be seen in one of the photographs of a livestock building at Kensham Farm, adjoining the farmhouse, where the tiled roof had failed and the tenant farmers had made a temporary repair with a patch of thatch – capable perhaps of keeping out much of the rain, but not a satisfactory long-term repair.

The lack of facilities in many farmhouses at that time may seem astonishing by modern standards, with many young folk today finding it hard to believe that right up to the mid-1950s there were houses with no indoor lavatory, only an earth closet out of doors. At Kensham Farm the members of the Bird family needing to relieve themselves had to go out of the external back door of the farmhouse into the garden to the earth closet, which was built with bricks adjoining the back of the farmhouse. In this privy there was a wooden seat with one hole in it, but no light other than daylight or the light of a hurricane lamp (burning paraffin through its wick) which they had to carry if it was after dark – and of course there was no heat, although having no water there was no danger of frost damage. From time to time the menfolk had to empty the earth closet – a system under which the vegetable garden thrived, a significant benefit of this recycled compost which increased the fertility of the soil.

Kensham Farm was described in the Parmoor Estate Sale particulars as having a 'soft water supply', which meant that the rainwater off the gutters of the farmhouse was saved into a brick-built chamber like a well, below ground in the courtyard of the farmhouse, for drinking water. At that time the Bird family womenfolk had to pump up this saved rainwater with a semi-rotary pump in the kitchen, using a backwards and forwards motion on a large handle to fill a bucket to be used for all cooking, washing and laundry. When my father arranged for the Marlow Water Company to install a main water supply to the farm, the Bird family womenfolk found the installation of one tap for cold water in the kitchen to be a remarkable modern improvement which eased their daily living very significantly.

A pond on Cadmore End Common in front of the farmhouse was the water supply for any livestock on the farm. This pond, from which the cattle had to drink, had been a serious health risk, since the pond was infected with Johne's disease. This is an unpleasant disease caused by the bacterium M. paratuberculosis which embeds itself in the lower part of the small intestine, thus preventing nutrient absorption and resulting in weight loss. There is no known cure for Johne's disease; infected cattle lose weight to the extent that all the ribs on their body can be seen in the weeks before they die from the condition, and when the microbe is excreted into the pond while infected animals are drinking it adds to the risk for the remainder of the herd.

The great benefit from the new piped water supply to all the fields was that the cattle could now drink clean mains water. There were no more new cases of Johne's Disease after the installation of main water, proving that the pond must have been infected with this unpleasant disease.

There was no electricity in the farmhouse, so that lighting had to be with either paraffin lamps or candles. Heating was with open fires, and for cooking there was a kitchen range, shown on the left of the photo of the original kitchen, which was normally fuelled by logs from the farm. The modern equivalent of a kitchen range would be an Aga cooker. The kitchen range in the Kensham Farmhouse had a compartment that could be filled with water so that hot water could be drawn from a brass tap on the front when it was going well.

When my father had laid on the single-phase mains electricity supply it had to end at a pole by the side of the farmhouse, since the Bird brothers declined my father's offer to have it connected to the farmhouse. The Bird brothers could see the advantages of electricity, but they also feared the cost of it, and so never agreed to its installation in the farmhouse on the grounds that they had always managed without electricity in the past.

Another improvement which my father commissioned for the farmhouse in the early 1950s was the rebuilding of one of the two front gables of the farmhouse which had been leaning dangerously, with the possibility of collapse. He had taken me with him to meet the local builder on site for an inspection of the dangerous gable, prior to its rebuilding with re-used handmade bricks to match the original. However, installation of electricity to the interior of the farmhouse, and a hot and cold water plumbing system with indoor lavatory, had to wait until Michaelmas 1955 when my late dear wife Alison and I took over management of Kensham Farm, six weeks after our wedding at St Giles, Ashtead.

On one occasion when my father took me on a visit to Kensham Farm in the late winter we saw Jack Bird preparing the daily ration of mangelwurzels from a clamp for the cows' feed. Mangelwurzels, often called mangolds, were a root crop not dissimilar to sugar beet. If the mangolds became frosted they were spoiled, and it was therefore necessary, after lifting the crop in the Autumn, to put them into a clamp, shown in one of the photographs. A clamp is a long heap of the mangolds covered with straw to exclude frost, with soil over the straw to prevent the straw from being blown away by the wind. When the mangolds were needed in the late winter for feeding the cattle each day's ration had to be taken out of the clamp by hand and then put through a root cutter, which sliced them ready for the cows to eat. A huge amount of manual labour was involved in feeding the cows by this method, and the mangolds themselves had a relatively poor nutrient level.

It was therefore not surprising that in the post-World War II days considerable research into livestock nutrition was taking place. Professor Bobby Boutflour CBE, who was Principal of the Royal Agricultural College, a role to which he had first been appointed in 1931, who was said to have been, 'The most capable, colourful and forceful character that ever adorned the fields of agriculture'. He was one of the fiercest opponents of mangolds as a feed for cattle, telling us students that they contained very little other than water. The Professor's view of cattle nutrition was that dairy cows with the potential from their breeding to give good milk yields should be given sufficient concentrates, that is prepared feeds supplied in a bag, to achieve their potential. Such concentrates were based on barley and oats with protein, often in the form of soya bean meal and fishmeal or meat and bone meal, with added trace elements.

At that same time the National Agricultural Advisory Service, 'NAAS', was active, giving good advice free of charge to all farmers choosing to make use of the service. The chief Advisory Officer for Buckinghamshire at that time was John Stubbs who, at my request, made a most useful advisory visit to us at Kensham Farm, in our early days of our farming with a dairy herd. He had written a book entitled 'Forage Farming', forcefully putting forward the view that it was uneconomic to feed dairy cattle too much concentrates, and that instead their diet in the summer months should be just grass, and that the winter feeds should be based on silage and kale.

So those were good times for the farming industry, many Universities, Colleges, and Government Demonstration Farms were carrying out research into livestock nutrition, and we had these two vociferous experts each recommending opposing methods of feeding cattle. Most of us farmers were able to listen to both extremes and manage our own farms along a course somewhere between those extremes.

Jack and Frank Bird at Kensham Farm

The Bird brothers used two ancient Fordson Standard tractors which dated from the 1930s, one on rubber tyres, and the other which was used for ploughing had wheels with iron spade lugs instead of tyres. One of the photographs shows Frank Bird ploughing with it, drawing a two-furrow plough. At Jack and Frank Bird's livestock and deadstock sale at which they sold all their cattle and poultry and farming equipment we bought that Fordson on spade lugs and used it as our second tractor when we started farming here. These tractors ran on tractor vapourising oil (TVO) which made a lovely smell when the engine was hot and running, but it was not sufficiently volatile to start a cold engine. For that reason there was also a small tank of petrol for starting the tractor, with a changeover fuel tap that had to be changed to the TVO setting once the engine was warm and running smoothly. There was no battery or electrical system on the tractor, only a Magneto, so that starting had to be by swinging the starting handle with the fuel changeover tap on the petrol setting. Once the engine was sufficiently warmed it would be switched over to run on the TVO, which was a cheaper fuel than petrol.

I remember another occasion in 1948 when Frank Bird told me about a new garden centre which had been set up at the nearby village of Studley Green. Before that time plants for gardeners were normally supplied by a nurseryman in a smaller way of business. Present day garden centres, selling plants that had been grown elsewhere and then sold at a branch of a chain of garden centres, together with tools and fertiliser and other requisites for the garden, did not exist at that time.

On another occasion, when Alison and I were engaged to be married, we were on a short holiday by the South Coast and spent a day walking on the South Downs near Steyning. There we saw a blacksmith and wheelwright putting the iron tyre on a cartwheel that he was repairing. The wheelwright had to forge the iron tyre to be marginally smaller in circumference than the wooden cartwheel to which it had to be fitted, and then heat it up to become red hot and thus expand it to become a slightly greater circumference than the cartwheel. Then the wheelwright with several helpers lowered the red-hot iron tyre onto the wheel which had been laid flat outside the Smithy and positioned it with hammers and levered hooks called devil's claws to become a good fit on the cartwheel. The several helpers immediately had to tip buckets of cold water over the hot iron tyre to cool it off before the wood of the cartwheel caught fire. The wheelwright had to be a highly skilled craftsman to make cartwheels which were expected to have a long life carrying heavy loads over rough ground. It was interesting that even in 1954 such cartwheels were still in use.

Sacrewell – William Scott Abbott's farm of 600 acres

It had been in the Summer of 1953 that I made the decision that even though my father and grandfather were solicitors practicing in their own family firm in Lincoln's Inn, London, and that I had started on that route by serving Articles with Jaques & Co in the City, a firm in which the senior partner was one of my father's friends, nevertheless that was not the life and profession that I wished to follow. I therefore approached Sandy Lee, a friend of our family who farmed in Sussex to offer to work as a harvest casual – that was a good start, resulting in an introduction from him to William Scott Abbott of Sacrewell near Peterborough where Sandy himself and his son had also learned the farming trade.

My year learning the farming trade from William Scott Abbott, and living in the farmhouse at Sacrewell with him and his wife Mary was an immensely formative year for me, having a significant influence on my future career. It was not only the way he farmed, and the way he thought and his philosophy of life, but also in practical terms it was his emphasis on efficiency and good planning leading to the farm being profitable - a necessity that most farmers who are still farming have learned, whereas those who never learned it are not still farming.

William Abbott had trained as an engineer, then on war service drove an ambulance in World War I before taking over from his father's tenant at Sacrewell Farm in 1917, which he ran with great success for 42 years until his death in 1959. During that time he regularly took on farm pupils as trainees. In the year when I worked there, from Autumn 1953 until after harvest in 1954, I was one of three trainees within a total staff of 27. By today's standards it would be unusual for a farm of 600 acres to have more than one farmworker to assist the farmer, but in 1953 the post-war drive for mechanisation, for large machines to replace manual work, had hardly started.

In those days Sacrewell was a mixed farm, growing wheat, barley, oats, mustard for Colman's, and sugar beet for the processing factory at Peterborough, which had been built during the 1920s but was subsequently demolished in 1991. On the livestock side, there was a flock of free-range pullets for egg production, and a dairy herd of Jersey cattle, from which the milk was bottled in the farm dairy and then retailed in local villages and the town of Stamford, for which the farm ran three of its own delivery vans. The herd was normally milked by two herdsmen with one trainee. There were three bulls, and the female calves were all reared as replacements, with any not required being sold for milk production in herds elsewhere. Milking started at 4.0 am, so that the bottled milk could be delivered to homes in Stamford in time for breakfast.

Work such as hedging and ditching that would always be carried out with machines nowadays was done in the traditional way by hand. One of the experienced farm workers kept his hedge slasher blade as sharp as a razor. It had a haft about 3 feet long, and he trimmed several miles of hedge neatly with it each winter. Another of the farm workers was expert at digging trenches, often the full length of a field to an even and gradual fall towards the ditch, prior to laying land drains in it and starting the backfill with shingle. There were still seven working horses, and William Abbott himself would always be at the cart shed at 7.30 am when the foreman, Mr Waller, allocated jobs for the day to the arable and general farm workers.

The first root crop to harvest in the autumn was potatoes. I was generally allocated the Suffolk Punch draft horse, Punch, to harness him up and put him into the shafts of the tumbril cart prior to working with him all day loading baskets of potatoes, which had been filled by womenfolk from the village picking them up behind a tractor drawn simple type of potato harvester, which lifted the crop, shaking soil off on its moving elevator, and then dropped them onto the soil in rows.

Sugar beet was another crop that involved much handwork. It was seeded with a horse drawn seed drill, then the emerging crop was hoed between the rows with a horse drawn hoe. The next stage of hoeing within the rows, to give each individual plant a suitable growing space before the next plant, was handwork with a hoe. Just like gardening, but on a field scale. During the sugar beet harvest in the autumn a tractor-drawn lifter would loosen the root of the sugar beet by lifting it and leaving it on the soil. Then the hand work started, with the first worker knocking the beet together to dislodge any soil prior to cutting off the green tops with a billhook and dropping them back onto the soil. At that stage I was normally one of three with horse drawn tumbril carts whose job was to lift the beet off the soil with a special hand fork (that had rounded steel at the ends of its tines, so that they would not stick into the beet) to load the tumbril cart. We would then lead the horse to the clamp at the side of the field, where we had to pull out a retaining bolt so that the tumbril cart would tip the beet onto a large heap near the field entrance gate. The beet would subsequently be loaded by hand into the farm's own lorry of 4 tonnes capacity, using a Lister moving belt elevator with its own small petrol engine, for transport to the sugar beet factory near Peterborough. At the factory the sugar beet was processed, the ultimate product being packets of sugar for sale in grocery shops. Nowadays, more of the sugar sold in Supermarkets originates from sugar cane grown in warm overseas countries.

Following William Abbott's death on 24th November 1959 his widow, Mary, embarked on what needed to be done to form the Trust they had intended to establish in their lifetime together. The Trust was formed in 1964 under the umbrella of the Royal Agricultural Society of England (RASE). It remains autonomous today with its own Board of Trustees who continue the vision and work of William and Mary Abbott. Sacrewell is now used as an educational visitor centre with a large café and playground – a popular place for Midland families with children to visit for a day out on the farm.

Cereal growing changes since the time when I worked at Sacrewell have been dramatic, resulting in a far smaller work force on the farms. Few crop protection products had been developed in the early 1950s, so the first task was to ensure that an emerging crop would not be choked out by weeds such as couch grass. Control of couch grass, to be effective in destroying its extensive root system with which axillary buds can grow horizontally and then form new shoots, involved firstly cultivating the field with a light cultivator, then using a chain harrow to roll up the root rhizomes which had been pulled to the surface into clumps which were then destroyed by workers with pitchforks making small bonfires over the whole field. Nowadays, one herbicide treatment of glyphosate on a field of stubble will eliminate the couch grass, by treating any shoots or leaves above the soil with the spray treatment translocating to the whole of the couch grass system of rhizomes below ground level.

In earlier times a crop rotation, such as the Norfolk Four-Course Rotation of wheat, turnips, barley, clover, was used to control weeds, since any weeds that flourished in the wheat crop were then destroyed by hand hoeing in the following crop of turnips. There was the further benefit that *Rhizobial* bacteria in clover root nodules fixes atmospheric nitrogen which will be available to the succeeding wheat crop. The principle of the rotation was thus to maintain fertility for the wheat from which flour for making bread, the staple food for human consumption, could be milled either in a water mill or in a windmill. At Kensham Farms many of our fields now grow continuous wheat, year after year. We are able to maintain fertility by the use of appropriate fertilizer, sometimes supplemented with farmyard manure from neighbouring farms, or with sewage sludge, and keep the field free from weeds by the use of appropriate herbicides.

The final stage of growing a cereal crop is harvest, when the crop is cut and the grain is separated from the straw. It is the harvesting process where the greatest changes have taken place during the years since the early 1950s while I have been farming.

Harvesting with a Self-Binder

When I worked for William Abbott all the cereal crops, of wheat, barley and oats were harvested with a tractor drawn self-binder. Under this system the crop was cut with the self-binder which then tied the stems of the crop, with the grains still inside the ears, into sheaves, leaving the sheaves on the ground. The next step was for farm workers to pick up all the sheaves, making them into stooks generally of twelve sheaves in each stook, stacked so that the heads were always at the top where wind and sunshine on the ears would complete the ripening and drying process. Some two or three weeks later the stooks would be loaded by hand with pitchforks onto a cart, to take them back to the stackyard to offload them, prior to building them into a rick. The rick would then have to be thatched with long stems of straw to make it rainproof. By then the month would probably be September.

Later in the autumn or winter the grain would be sold for despatch off the farm in railway sacks, holding 2.25 cwt (approximately 114 kilogrammes) of wheat in each sack. But it first had to be threshed, the process of knocking the grains of wheat out of the ear. This was done with a threshing machine, drawn and then powered by a steam traction engine (shown in photographs in the first chapter). At that stage the sheaves of wheat had to be manually handled yet again when the thatch was stripped off the stack to open it up for threshing day, when one worker would throw the sheaves up to the top of the threshing machine to be caught by two workers who would cut off the twine binding each sheaf before feeding them evenly onto the drum of the threshing mechanism. Thus the wheat grains, either in the stems of the plants before threshing or as grain in railway sacks after threshing, would have been picked up and put down by human muscle not less than nine times, sometimes more, before the wheat grown on the farm could be delivered to the mill for the first stage of the beadmaking process.

Harvesting with a Combine Harvester

At Kensham Farms we now have a John Deere combine harvester which normally harvests grain at the rate of over 45 tonnes per hour, all the handling being in bulk, so that at no stage is human muscle used for lifting any of the grain. On a favourable sunny day, with a clean crop of wheat in a large field this rate of harvesting is sometimes up to 60 tonnes per hour, that is 1 tonne of wheat harvested every minute.

The silo or grain store has to be capable of receiving grain at this rate, with all its elevators and conveyors being capable of working at this speed. In Britain, with our known weather systems, wheat growers always have to make provision for drying the grain in those summers when there is insufficient sunshine. A few farmers rely on grain drying and storage cooperatives, where the grain leaves the farm perhaps with a moisture content of 17% or more to be dried in an industrial scale grain dryer prior to storage and sale. However, we and most other farmers have our own grain dryers, in our case a Scandinavian Svegma Drier with a capacity of removing 5% of moisture from wheat at the rate of 46 tonnes per hour. Thus, with the modern self-propelled combine harvester, a system first developed in 1911 in California by the Holt Manufacturing Company, the earlier need for a large workforce of strong workers at harvest time has been eliminated. The present-day requirement is for a few farm workers who have great skill and mechanical ability.

Retirement of Kensham Farm Tenants

In 1954, while I was still studying agriculture at the Royal Agricultural College, Cirencester, Jack and Frank Bird gave in their notice to my father to end their tenancy of Kensham Farm, with a view to retiring and moving out at Michaelmas 1954. That was when my father asked me if I wanted to have a go at running the farm after I had left college, in the capacity of farm manager to the family company, The Sonning Land Company Limited. This was a family company founded in 1924 by my father and grandfather with Frederick Fisher, a shipbroker. The Company had bought Charvil Farm with land adjoining the River Thames in 1924, and after that bought various small houses on newly developed land in outer London. In those early days between the two world wars the Company had not prospered, so that when Mr Fisher wanted to pull out my father was able to buy his shareholding in the Company for a modest figure. My father took the risk that it might have been a bad investment, since the Company's survival was in great doubt in the difficult times of the 1930s.

I showed my enthusiasm to my father for his offer to employ me, since I had previously been hoping to get a management job somewhere in the farming industry not connected with my family's interests. My father then asked the Bird brothers if they would be prepared to continue as tenant farmers until Michaelmas 1955, rather than 1954, so that I could take over Kensham Farm after leaving the Royal Agricultural College and marrying my Alison to whom I was already engaged. When my father promised to waive any dilapidations that would have been due from the Bird brothers, and to take a relaxed attitude over the rent of £77 pa for the last year that otherwise would have been due, the Bird brothers agreed and duly vacated Kensham Farm on Michaelmas day 1955, the 29th of September.

Security of Food Supply in Britain from 1947 to 2020

The Agriculture Act 1947
The background to the Agriculture Act 1947 passed by Tom Williams, Minister for Agriculture in the post-World War II Labour Government, was the continuing necessity for food rationing. Tom Williams introduced the Agriculture Bill by explaining that his objective was:-

> *"to promote a healthy and efficient agriculture capable of producing that part of the nation's food which is required from home sources at the lowest price consistent with the provision of adequate remuneration and decent living conditions for farmers and workers, with a reasonable return on capital invested".*

The Members of Parliament in 1947 had all lived through the Second World War and so had first-hand experience of real shortage of food, eked out through the system of ration books which continued in use until 1954, and so were pleased to pass this new Agriculture Act in the knowledge that never again would the people of Britain face such a shortage of food.

'Health and Harmony' Consultation Paper, prior to enacting the Agriculture Act 2020
I think it quite extraordinary that many of our present-day leaders in the political sphere, and broadcasters on television, in the present discussion prior to enacting The Agriculture Act 2020, have forgotten past history of the need to feed the British people. Such is their lack of recognition of the importance of food security, that when they are considering questions of the environment and wild animals, birds, bees and butterflies, and other important issues like quality of water, they forget that food is necessary for the human race, but instead talk only

about rewilding and measures to look after wildlife rather than the need to grow food for the human race and in particular the British population.

In February 2018 Defra published two important reports concerning the future of farming in Britain in a post-Brexit world. The first of these was published by Defra in conjunction with the Government Statistical Service and is entitled: -

'The Future Farming and Environment Evidence Compendium'

This evidence compendium is the most thorough, clear, well presented and accurate Government report on present day farm costs, returns and profitability that I have ever read in the 67 years in which I have been actively involved in the farming industry.

In the very same month, February 2018, Defra published another report, of similar length, 64 pages long (as compared to the Evidence Compendium of 68 pages). This second report is entitled: -

'Health and Harmony; The future for food, farming and the environment in a Green Brexit'.

and has a Foreword on its fifth page by The Rt Hon Michael Gove MP, who at that time was Minister for Defra. I consider this title, 'Health and Harmony' to be an absurd title for a document which is supposed to set out the way forward for British farming following Britain's exit from the European Union.

The policy put forward in this 'Health & Harmony' consultation report is effectively to ignore the production of food in favour of other worthwhile aims such as care of the environment. It would have destroyed the aim of the Agriculture Act 1947, which was to ensure food security for the people of Britain.

I feel that the determination of the Secretary of State for Defra to cut out direct support for farms producing food, described in the 'Health and Harmony' policy document, was an act of political folly. Furthermore, the notion held by the Secretary of State, that provision of a secure supply of good quality food is not a 'public good', is so false as to be ludicrous.

If the author of the 'Health and Harmony' report had studied and taken note of the statistics set out so well in Defra's own publication, the Evidence Compendium, then he would know that a very high proportion of British farms as we know them today would no longer provide a living wage for the farmer and his family. Those farmers would be forced out of business if the initial 'Health & Harmony' proposals become Britain's new agricultural policy.

Defra 'Farmer Engagement Meeting' at Kensham Farm

We were pleased that the 'Health and Harmony' paper was a consultation paper, so that we were able to arrange one of the 'Farmer Engagement Meetings' here at Kensham Farm in our dining room on 28th February 2018 at which there were three Defra officers, headed by Tony Pike, and 15 of us farmers with interests in arable, agronomy, dairy, smaller farm businesses, farm shops and other diversifications. We were pleased that our deliberations were reported accurately in the Defra summary of our consultation meeting.

We felt that the Minister for Defra should realise that farmers look after around 70% of the land in Britain, and that if they are threatened with bankruptcy the first thing that they will be forced to curtail is their care of the rural environment, the very thing which the Minister purports to be keen on preserving. In the response by myself and my family to the 'Health & Harmony' consultation, we set out the view that the Agriculture Bill should achieve two principal aims, which are for Government to: -

- **Encourage and support lowland farmers to produce food and care for the environment as a by-product.**
- **Encourage and support farmers with Open Access Land (designated by Defra as Less Favoured Areas) to look after the landscape and environment, and to produce food as a by-product.**

The Agriculture Act 2020

This new Agriculture Act was formulated over a period of almost three years. It started with the impractical Health & Harmony Consultation Paper described above, then the next stage was when Defra produced the Agriculture Bill, in which food production was still not treated as the main aim. Any Bill must go through the process of the First and Second Readings in the House of Commons, followed by the Committee Stage, the Report Stage, and finally the Third Reading, then a similar sequence in the House of Lords. In this case of the Agriculture Bill their Lordships suggested most useful amendments concerning the importance of food. During those stages the NFU, led vigorously by its first ever female President, Minette Batters, pressed hard for the importance of growing food to be recognised.

The Agriculture Act 2020 was enacted by receiving Royal Assent, appropriately on Remembrance Day, the 11th of November 2020, the day on which it became law. By then many of the improvements suggested in the House of Lords following considerable NFU pressure were included, these being the importance of Food Security, and the need to encourage food production from British farms. The Act also gives Government powers to intervene if there were to be unexpected market disruption, and a Trade and Agriculture Commission has been formed to scrutinise future Free Trade Agreements. This will provide MPs with a report as to how each Agreement with different nations will impact British farms, including whether imports would be of such low quality that they would have been illegal if produced in Britain.

My final comment on this subject of Government support for growing food, is that direct farm support (funded in recent years by the EU for Britain, and with the USA and most other countries in the world other than New Zealand having their own support systems) enables ordinary citizens to buy food, with a certainty of availability, at a price that is less than the cost of production. Direct farm support for food production is therefore really a subsidy that provides cheap food for everyone. There is also the further advantage that the countryside and wildlife can continue to be looked after by farmers as a by-product of the production of this food.

Food Security

My articles entitled *'On the land'*, now reproduced within Chapter 5, were written between 2007 and 2021 for The Clarion, which is published by the Lane End Parish Council magazine. Some of these articles have been written to draw attention to the food shortages that threatened Britain during two World Wars. It seems that, yet again, some politicians appear to have forgotten the importance of food security. I realise that no active politician in 2020 is old enough to remember the rationing of food in either of the two World Wars.

The Edgley Family at Kensham Farm

In the next chapter, I will describe the sixty-five years of my own experience of farming Kensham Farm, and then in Chapter 4 those topics allied to the farm, but not specifically part of the growing and production of food from our own land, with which I have been involved over these years.

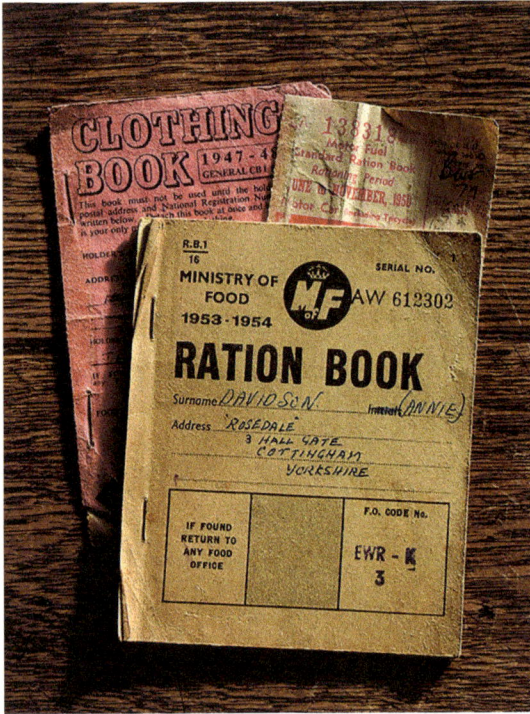

DIG FOR VICTORY

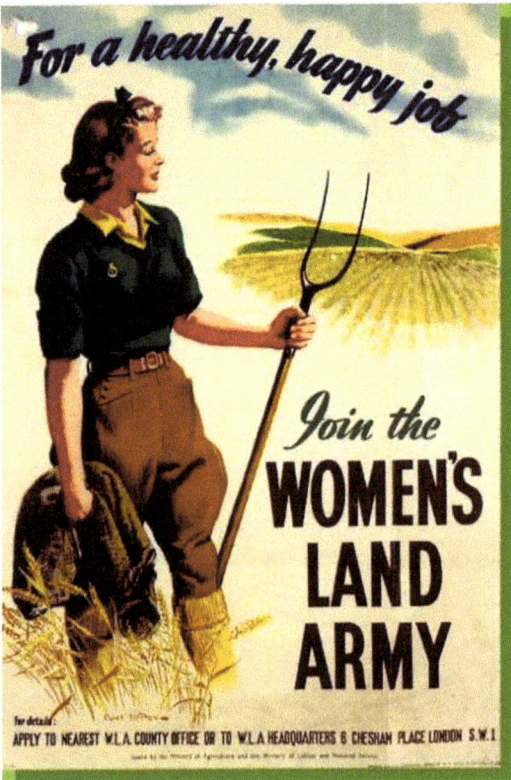

For a healthy, happy job

Join the WOMEN'S LAND ARMY

APPLY TO NEAREST W.L.A. COUNTY OFFICE OR TO W.L.A HEADQUARTERS 6 CHESHAM PLACE LONDON S.W.1

All UK citizens were issued with ration books containing coupons which the retailer had to tear out and retain against sales of most foods and clothing. Rationing started in 1940 and ended in 1954.

The Dig for Victory campaign encouraged those who had gardens, or an allotment, to grow their own vegetables.

This recruiting poster was widely displayed in the early years of World War 11 to persuade young women, who were not serving in the forces or other essential war work, to volunteer to work on the land. By Autumn 1941 20,000 women had already volunteered to become Land Girls, around a quarter of whom were employed in dairy work.

BUCKINGHAMSHIRE - OXFORDSHIRE BORDERS

On the hills between High Wycombe and Henley

In the heart of some of the most beautiful country in the
Home Counties, close to Turville, Fingest and Lane End

High Wycombe 6½ miles, Marlow 5½ miles, Henley 10 miles and Stokenchurch 2 miles

Particulars, Plan and Conditions of Sale of the

CADMORE END ESTATE

in the Parishes of
STOKENCHURCH and FINGEST

About 686 acres

comprising

POUND FARM — 141 acres   RACKLEYS FARM — 54 acres
KENSHAM FARM — 102 acres   WATERCROFT FARM — 107 acres
DELLS FARM — 116 acres   GIBBONS FARM — 86 acres

Also Smallholdings, Building Sites and 17 Cottages with main
electricity and water already installed or available in many cases

Which will be offered for sale by auction in lots as described in the following
Particulars (unless previously sold privately) by

MESSRS.

JAMES STYLES & WHITLOCK and JOHN D. WOOD & CO.

(acting in conjunction)

At the GUILDHALL, HIGH WYCOMBE.
On Friday the 29th day of JUNE, 1945 at 3.0 p.m.

Vendor's Solicitors: Messrs. MORRELL, PEEL & GAMLEN, 1 St. Giles, Oxford. Phone, 2468.
Resident Agent: Mr. G. SHERWIN, Estate Office, Parmoor, Henley-on-Thames. Phone, Lane End 215.
Auctioneers' Offices: Messrs. JAMES STYLES & WHITLOCK, 16 King Edward St., Oxford. Phone, 4637
Also at St. James' Place, London; Chipping Norton; Rugby and Birmingham.
Messrs. JOHN D. WOOD & CO., 23 Berkeley Square, W.1. Phone, Mayfair 6341.

Lot 9
(Coloured Blue on Plan.)

KENSHAM FARM.

An Attractive Holding of

102a. 1r. 11p.

with a

MOST PICTURESQUE OLD GABLED FLINT AND TILED FARMHOUSE

occupying an almost perfect setting on Cadmore End Common across which it is approached by a gravelled road.

The House has great possibilities for **conversion into a Small Country House** and stands in a small low walled garden overlooking a tree shaded pond.

It contains : living room, kitchen, back kitchen, scullery and old dairy on the ground floor and 3 bedrooms and a boxroom above. E.C. outside.

Soft Water supply.

Electricity and Main Water are available at the entrance to Cadmore End Common, about 250 yards away.

THE FARM BUILDINGS

comprise

timber built and tiled tractor house, brick and tiled barn with weather boarded upper part, 3-stall stable and 2 boxes, cowhouse for 5 and another for 4. Another barn of flint and weather boarding, with tiled roof, calving box and 3 pigstyes.

NOTE.—The tenant claims the timber sheds and the iron shelter in the yard.

Form of Agreement

AN AGREEMENT made the 29th day of *July* 1945, between THE RIGHT HONOURABLE ALFRED HENRY SEDDON BARON PARMOOR OF FRIETH

(hereinafter called "the Vendor") by his agents mentioned below of the one part and *Richard Bernard Smith, Wycombe Road, Stokenchurch, Bucks*

(hereinafter called "the Purchaser(s)" by his (their) agents mentioned below of the other part WHEREBY it is agreed that the Vendor shall sell and the Purchaser(s) shall purchase Lot(s) *18, 20 & 30* described in the above Particulars at the price of *Three thousand five hundred & ninety* (£3,590) (independently of any valuation money) subject to the foregoing Special Conditions of Sale and the Law Society's Conditions of Sale (1934 edition).

AS WITNESS the hands of the parties hereto or their agents.

	£	s.	d.
Purchase money	3590	0	0
Less Deposit	359	0	0
Balance	3231	0	0
Valuation money (if any)	nil		
Total	3231	0	0

Abstract of title to be sent to

As Stakeholder(s) we hereby acknowledge the receipt of the above-mentioned deposit this

day of *29th July* 1945.

Opening page of the sale catalogue of Lord Parmoor's sale of 6 farms and 17 cottages comprising 686 acres in total, of his Cadmore End Estate on 29th June 1945.

Particulars of Kensham Farm shown in the Parmoor Estate sale catalogue.

Form of agreement on the final page of the Estate sale catalogue of the Parmoor Estate sale on 29th June 1945 of Lots 18, 20 and 30 sold to Richard Bernard Smith for £3,590.

OBITUARY
MRS MARY ABBOTT
Farming and care of the old

A correspondent writes:

Mrs Mary Abbott who, with her late husband, William Scott Abbott, devoted her life to farming and the care of the elderly, has died at the age of 92. She was the youngest child of the Rev Townsend Powell and his wife Mary. Her father died in Switzerland, when she was nine years old. From then until her marriage, apart from an adventurous journey to China and the Far East, followed by land work during the First World War, she lived with her mother and elder sister.

In 1923 she married William Scott Abbott of Sacrewell Farm, Thornhaugh, near Peterborough. They had no children, but their 36 years together were devoted to their common interests in the farm at Sacrewell and farming ventures in Yorkshire and Herefordshire, on which they worked in closest partnership. Their efforts to rejuvenate a run-down block of hill farms in Bransdale, on the North Yorkshire Moors, between 1940 and 1958, was a useful project, which gave them great interest, but little financial gain.

At the Dulas Court Estate, between Hereford and Abergavenny, which was bought in 1953, they made great sacrifices in running the Court as a home for the elderly, but derived much interest from the forestry and farming of the estate.

At Sacrewell the Abbotts' farming was always of an extremely high and progressive standard and this provided a base to support their more altruistic ventures. Sacrewell has also, over the years, been a first class grounding for a great many farm pupils.

William Abbott died in 1959, aged 70, for the last two years of his life nursed by his wife through advancing stages of sclerosis. As her husband's executrix and sole beneficiary, Mary Abbott set about the administration of his estate and in 1964 the Trustees of the Royal Agricultural Society of England accepted the Trusteeship of the William Scott Abbott Trust. The settlement consisted of Sacrewell's 575 freehold acres, complete with live and dead stock, as a Charitable Trust in memory of her husband and their work together.

The objectives of the Trust are, broadly speaking, Agricultural Education and Research. The Trust, now entering its 14th year, has already, both in work at Sacrewell—most notably through the provision of facilities to the Processors' & Growers' Research Organisation (PGRO)—and through its close associations with the Royal Agricultural Society, achieved much of real value to farming and food production and its potential for the future is very considerable.

In 1968 Mary Abbott gave the house and grounds at Dulas Court to the Trustees of the Musicians' Benevolent Fund, who have, with care and much effort, developed it into a fine home for retired musicians.

Obituary in The Times of Mary Abbott, William Scott Abbott's widow.

Aerial photograph of the Sacrewell Farm House and farmstead in 1948. It will be seen that there are no corn ricks in the stackyard, since the photo was taken in the summer, after threshing of the previous year's ricks and before harvesting the 1948 crops.

SACREWELL 1948

Corn ricks in the Sacrewell stackyard, photo by the author in Autumn 1953. These were the ricks threshed in February 1954, at the time when the author was one of the workers in the threshing gang.

The Mill House at Sacrewell, with the stables on the right and the end of the wagon hovel on the left. Some horse-drawn harrows are stacked against the wall.

The Stables at Sacrewell in which the seven working horses were stalled in the 1950s.

The Wagon Hovel at Sacrewell. In 1954 it housed two small Ferguson tractors, as well as the horse drawn tumbril carts. All the general farm workers, including the author in 1953/54, assembled in front of the wagon hovel at 7.30am every morning, to be allocated the day's work by the foreman, Mr Waller, overseen by William Abbott.

Bill Ayres, one of the general farm workers at Sacrewell in 1954, with one of the seven working horses and the seed drill that was in current use at that time.

Ron Simpson, the Head Carter at Sacrewell in 1954. It was the Head Carter's responsibility to run the stables and be responsible for the health and fitness to work of the seven working horses.

Topping sugar beet on to cart to be loaded direct on rail

This photo, taken from a booklet published by William Abbott on the farm work at Sacrewell at the end of the 1940s, shows one of the tumbril carts being used to cart lifted sugar beet from the field to a temporary heap by a hard road or track.

Photograph from the Sacrewell archives of the oat harvest in progress. The Standard Fordson tractor is being driven in the fields by a lad, enabling the tractor driver to stack the sheaves being tossed up with pitchforks by two men and one elderly woman.

Harvesting with tractor drawn self-binder in the 1940s. Photograph by kind permission of Sir Edward Dashwood Bt, from the West Wycombe Estate archives. The Parish Church of St. Lawrence, West Wycombe can be seen on top of West Wycombe Hill in the background. This harvesting method was essentially the same as that photographed by my father thirty years earlier, but with the self-binder being drawn by a tractor instead of horses.

Trouble with the Standard Fordson tractor. This tractor was on spade lug iron wheels rather than tyres, with the sails of the self-binder just visible on the left. A gun can be seen leaning against the rear mudguard of the tractor - this was probably a 12-bore shotgun, highly likely to be already loaded for the control of rabbits running out of the standing corn as it was being cut. In the 1940s farmers and others took a more cavalier approach to health and safety, with less control of firearms than would be acceptable in modern times.

Aerial photo from the archives of the Grassland Research institute, Hurley (1949-1992) of a row of 21 corn ricks at a Berkshire farm. The ricks would have been built adjoining a hard track, to allow for good access at threshing time. These ricks will have had a life of only a few months after harvest, before being threshed for the grain to be used or sold in the winter months.

Kensham Farm in 1948, one of the author's first photographs of the farm. His father's Standard 8hp car is parked in front. The pond, which was the only source of water for the cattle, was a health hazard, infected with Johne's disease.

The rear of the Kensham Farmhouse. At that time the dairy in which milk was cooled prior to sale in churns was the wooden shed in the courtyard, and two of the milk buckets are hung on the fence. During the 1956 renovation of the farmhouse the courtyard was paved, and now provides the access for both kitchen door (centre of photo) and a new front door on the left.

The entrance track into the farmyard, with one of the Bird brothers' neatly thatched corn ricks in the background. The barn on the left was structurally beyond repair.

Ricks of corn in the stackyard. This photograph of the corn ricks built and thatched by Jack and Frank Bird in the stackyard at Kensham Farm was taken by the author when still a school boy in 1949. Few changes in harvesting methods had taken place in the preceding fifty years. However, the first combine harvesters were being developed and used in the 1930s in the United States and Canada. The purpose of the corn rick was to store the sheaves of corn before thrashing, while the grain in the seed head was still on the stalk of straw. These ricks were all thatched with wheat straw to make them watertight, even though the ricks had a life of only a few months.

Frank Bird ploughing with the 1930s Fordson tractor on spade lugs drawing a two furrow plough.

Jack Bird removing the dairy cows' ration of mangold wurzels for one day, prior to chopping them with a root cutter.

Roy Edgley discussing cropping with Frank Bird.

Roy Edgley discussing farming and family matters with Jack Bird.

View from Malin Hill towards Crow Field at Kensham Farm in 1948. This same view has been used by the author's grandson, Alex Nelms and his fiancee, Annaliese Brightwell, at the foot of the invitations to their wedding in January 2022, just prior to publication of this book.

Stack of hay made by the Bird brothers in summer 1955. At that time the best quality hay was made by tedding the mown grass to dry it after cutting. Then when sufficiently dry, after some days of sunshine, it would be swept up to a field stack, and then thatched. The author took over this stack at the Bird brothers' live and deadstock valuation at Michaelmas 1955. The quality of this hay turned out to be excellent, so that it fetched a good price to a hay and straw merchant. The merchant arranged for an experienced farm worker to cut the stack open with a hay knife, and then hand tie it into trusses which were sold for feeding racehorses at Newmarket.

The yard at Kensham Farm in 1955, with manure heap in the centre and the wooden dairy shed on the right. The yard is now surfaced with tarmac for car parking. In 1964 the farm building with the small gabled hatch on the right was replaced with the farm office, as an extension to the farmhouse.

The cost of a skilled tiler to renew any broken tiles would have been an unnecessary expense since a temporary repair could be made by thatching the leaking section with straw grown on the farm.

The author's father, Roy Edgley, with the author's spaniel, Greenwell, in the Kensham Farm stackyard, 1948.

The barn on the right was renovated in 1956, and then re-roofed with new trusses purpose made to a historic design, as our Kensham Farms Millennium Project in the year 2000.

North East elevation of the farmhouse, with vegetable garden in front in 1948.

The 'Privy', the only lavatory facility for the farmhouse, was the small lean-to on the right. The Bird family had to leave the house by the back door, and go the few yards through the vegetable garden to relieve themselves. Photographed in 1956, after the gable window and hall window below it had been renewed.

The kitchen at Kensham Farm prior to 1955, with the kitchen range on the left, wood-fired copper for clothes washing, centre, and brick and tile sink on the right.

The author's wife, Alison, inspecting a doorway out of the kitchen prior to the 1956 renovation.

Chapter 3

Bryan and Alison Edgley and their family
at Kensham Farm, 1955 to 2020

Section 1, Michaelmas 1955 to 1959 –   **The early years of mixed farming, with**
dairy, cereals, pigs and egg production

Alison and I were young when we married in August 1955; she was 22 and I was 23. Alison had taught Physical Education and Games at Surbiton High School after leaving Bedford College and I had completed my National Service and worked for a year in a solicitor's office in the city before the year at Sacrewell. I then studied agriculture on the short course at the Royal Agricultural College, Cirencester, which had a duration of one year. I and many other students on that course had already served in the Army for National Service, or experienced other occupations since leaving school.

Following our wedding we had a honeymoon of ten days or so in Scotland, travelling up there in the Ford van which I had recently bought for the new farming enterprise. When we returned, we had a few weeks to spare before taking over possession of Kensham Farm at Cadmore End, during which we took up my sister Diana's invitation to stay with her and Peter and their children at their house at Farnham Common.

In Diana and Peter's garden I constructed five pig huts, each about 7 foot by 5 foot in sections ready for the sows that we intended to keep when we started farming at Kensham. When completed, we had the sections transported on a lorry to take them the 10 or 12 miles to Kensham Farm. Alison and I looked forward with eagerness to the new adventure at Kensham Farm, which was to be our home together for nearly 65 years.

We knew that we would not be allowed to move into the farmhouse at Kensham Farm until it had been improved, since the Local Authority had placed a slum clearance order on it with effect from the date when our predecessors, the Bird family, vacated. We also knew that we had to carry out extensive repairs to the farmhouse, rather than rebuilding it, since it was a building of historical importance noted under the Town and Country Planning Act of 1947, and subsequently listed as a Grade II Listed Building in 1962. These repairs and improvements included the first installation of electricity, water system and interior WCs and sanitation that there had ever been in the farmhouse.

In 1954, during the time when I had been studying the technical side of farming at The Royal Agricultural College, my father had instructed a local firm of architects to prepare plans for building a pair of two-bedroom farm workers' cottages in the corner of one of the fields nearest to the village of Cadmore End. The Bird brothers readily agreed to cease farming that corner of the field, for we were on good terms with them following our agreement to take the land in hand to farm it ourselves when they retired.

Our architects instructed a good small local firm of builders, Dell Brothers of Lacey Green, to build the pair of new cottages, the total contract price being the order of £3,400. The then Lord of the Manor, Ronald Williamson, granted us access over a corner of Cadmore End Common from the B482 Marlow to Stokenchurch road to the proposed new cottages for which he made a nominal charge. Ownership of commons and the rights of the owners, who normally also hold the title of Lord of the Manor, was exceedingly vague before the Commons Registration Act of 1965 was introduced.

Alison and I moved into the first of the two new cottages as soon as it was ready for occupation in September of 1955. She had no previous experience of running a home and I had no previous experience of running a farm, although of course we both knew about these activities from earlier years, Alison's experience of housekeeping in her own parents' home, and my experience of farming under the tuition of William Scott Abbott and at the Royal Agricultural College. Initially I was Farm Manager to my father's company.

On Michaelmas Day, the 29th of September 1955, we attended Jack and Frank Bird's live and dead stock sale conducted by the local firm of auctioneers, Lawrence, Son and Laird of Marlow, and bought the Bird brothers' pre-war Standard Fordson on spade lug wheels and a four wheeled trailer and one or two smaller items of machinery, but most of the equipment was in poor condition. I particularly remember a friendly neighbouring farmer, Mr. Cluett of Pound Farm, coming up to me at that sale, which was attended by most of the neighbouring farmers, and introducing himself to me. I remember his greeting when he gave me his best wishes for our new venture and him saying that he hoped that I would be accepted locally, because he himself *"had come up from Somerset in 1910 and had only recently been accepted in Cadmore End"*.

In those days a village was a village, with a very strong identity of neighbours who all knew each other. Most villagers used to work at local trades, at which they were skilled after a lifetime's experience, with nearly all of them working within walking or bicycle distance of the place where they lived. The changes in the village of Cadmore End between 1955 and today are immense, but it is still a lovely village in which to live. Nowadays most of the residents are either retired or commute to jobs completely outside farming and the local community.

In the 1940s and early 1950s a farm of 100 acres was adequate for the farmer to earn sufficient money to live and bring up his family, and to employ a herdsman or general farm worker to assist him. A farm of 400 acres would have been considered a large farm and would probably have been an amalgamation of a few earlier small farms. This is a trend which has continued over subsequent years, since a mixed farm of 100 acres would no longer yield an adequate living for a single farmer, even without any staff to pay.

Most of the farms surrounding Kensham Farm on the Chiltern Hills used to have a dairy herd, albeit on a scale that would now hardly come into the category of serious farming. So it seemed that the best main enterprise for Kensham Farm would be a dairy herd, for which we chose the Friesian breed as being the highest yielding herd in which surplus bull calves also had a value for production of veal or beef.

At Kensham Farm our predecessors, the Bird brothers, had each milked their own small herds of four or five Shorthorn dairy cows by hand, each herd being independent of the other and in separate farm sheds. On most other farms at that time the earlier traditional system of keeping a dairy herd, which had been used ever since the time when milk maids sat on three legged stools milking cows by hand into a bucket, was still used. On that system each cow had her own allotted stall in a long cowshed, to which the cowman had to take the bucket type milking machine and connect it to a fixed vacuum pipe which ran along the length of the cowshed. The first mechanical milking machine had been invented in 1870, but it was several decades before design development had been improved enough to make them practical. Those earliest machines, some of which were painful for the dairy cows being milked, were superseded by the design that is still basically in use, whereby nature's way of the baby calf sucking the teat of its mother is replicated by the milking machine with four soft rubber cups fitting over the cow's four teats, pulsating under vacuum to provide the sucking action.

The yard and parlour system of housing and milking dairy cows, which was quite different to the old cowshed system, became standard practice after World War II. In this new system the milking machine was a fixture in the milking parlour which had an adjoining collecting yard. The cows waiting in the collecting yard prior to being milked had to walk into the parlour to be milked as soon as the preceding cow had been let out to walk back to the yard or field.

Our earliest farming at Kensham Farm followed the mixed farming pattern, which was the accepted norm in those days. Our intention had been to build a new yard and milking parlour for a dairy herd, and to grow some of the cereals for grinding with our own electrically powered hammermill as the main ingredient of the concentrated feed for the cows. We also intended to keep pigs and poultry as supplementary enterprises.

In those days the Ministry of Agriculture provided a fully funded advisory service for farmers known as the National Agricultural Advisory Service. With the NAAS farm buildings adviser I designed a new yard and parlour for the proposed dairy herd with sufficient fodder storage to last the winter season, all to be under a single roof in a new Crendon Concrete building which was to have a length of 75 feet, in 5 bays of 15 feet, and an overall width of 70 feet. It was divided into sections of the central barn with a high roof for self-feed silage for the dairy cows, a lean-to of 20 feet on one side for the storage of hay and equipment necessary for managing a dairy herd, and a lean-to 25 feet wide on the other side for the yard and milking parlour, suitable for a herd of 60 or 70 milking cows. Some of our neighbours thought I was crazy to be erecting such a large farm building at Kensham Farm.

I had behind me the valuable experience of having worked as one of the two assistant cowmen milking the Jersey dairy herd at Sacrewell. William and Mary Abbott came to see Alison and me at Kensham Farm during our first year – I still remember him looking at our thriving crop of oats which we had seeded after ploughing up a field of long-term permanent grass in the winter of 1955/56 and commenting that things would be alright when I had got all the other fields growing as well as that crop.

I established our dairy herd at Kensham Farm with an initial purchase of ten Friesian cows in milk from the dispersal sale of the Binorton herd in Lincolnshire. These cows were delivered to the farm just before the new yard and parlour building was completed, so in the early months I personally carried out all the milking in a system of hurdles in the field we know as Kiln Meadow. I used a bucket type milking machine with a petrol driven small vacuum pump. There was no roof and no electricity, and it was not the easiest job on a frosty morning to thaw out a frozen udder cloth in a bucket of cold water.

It may seem strange nowadays to milk cows out in the field, but the use of 'milking bails' had been first developed by Arthur Hosier in Hampshire in 1922 – he had wanted to make a portable outfit as a new way for milk to be produced cheaply, cleanly and naturally. The milking bail was really a small cowshed on wheels, without any floor, which was used in the field. Cows waiting to be milked were kept in a temporary coral on the pasture prior to going into the milking bail in which they were fed and milked in one of the six stalls. This had proved a successful system on free draining chalk land when it was followed by Rex Patterson who first used the system on a rented farm of 80 acres in Hampshire in 1927 and subsequently greatly expanded, to the extent that by 1942 he was farming more than 10,000 acres split into separate units, with a herdsman in charge of each milking bail unit. Careful records were kept in his central farm office which were used to foster a spirit of competition between units with the intention of making them more productive. However, the system became unmanageable, and was scaled right down after Rex Patterson's death in 1978. However, in his heyday, Rex Patterson and his milking bails had been widely reported in the farming press.

Turning back to the start of our own farming at Kensham Farm, by early Summer 1956 the new Crendon Concrete yard and our parlour building was up and in use, and the second farm cottage was completed and ready for use. At that stage we employed an experienced herdsman, Keith Burton, to milk the cows, so from that time I only had to milk our expanding herd when Keith had a weekend or day off. We also bought six in-pig sows and farrowed them in the pig huts that I had constructed in my sister's garden and started egg production on deep litter in one of the original barns.

At the same time that we were starting all this work on the farm, Alison was establishing our first home in No. 2 Kensham Cottages, the same cottage where my grandson Angus now lives. We were delighted when Alison found that she was pregnant, leading to the birth of our daughter Rosalind, now known to all her friends as Polly, in October 1956. We moved into the main farmhouse at Kensham Farm in Spring 1957 after its extensive renovation had been completed by Dell Brothers of Lacey Green.

My ambition in these early years, which has remained unaltered ever since, is probably the same ambition held by most farmers, to make sufficient profit to bring in improvements, balance the books and enough money to cover household living expenses. It is only by making a profit that a farm business is able to continue. When profit figures of limited companies are published, the managers and directors have already drawn their fees and salaries, whereas for self-employed activities or partnerships the profit figure is calculated before the proprietor or partners have drawn any return for their own management and manual work. None of our decisions were made on the basis that something 'might be fun' - all our enterprises had to be activities which were both suitable for the farm and which had the ability to yield an adequate profit margin. I had already learned from William Scott Abbott the enormous importance of keeping accurate cost accounts of all activities on the farm. There was no case then, nor has there been a case since, for a farmer to just farm and hope there would be a profit at the end of the year. Profit is something that should be at the core of every planning and cropping decision made on the farm.

On the machinery side, we bought a new Ferguson TE 20 diesel tractor, the *'Little*

Grey Fergie' model, that became so popular throughout the farming industry. It had been developed by Harry Ferguson who was granted a British patent in 1925 on the hydraulic valve which he had invented that enabled the implement being used and the tractor itself to become one unit, linked together by the three-point linkage. In this way the weight of the implement was transferred to the driving wheels of the tractor, increasing its traction before wheel slip occurred. Thus tractors no longer had to be as heavy as earlier tractors, to which the implement was joined to the tractor only with a drawbar pin. This did however mean that the light Ferguson TE20 tractor was best suited to farming with implements that had been specifically designed for it. On farms in many parts of the world this TE20 tractor replaced the draft horses which had previously been used for crop production.

At Kensham Farm we tried to cut corners on costs in the early years, one such economy being to use the second-hand four wheeled trailer that we had bought at the Bird Brothers' closing down sale, rather than to buy a new Ferguson 3 ton trailer with its pair of wheels positioned at the rear end of the trailer, thereby transferring much of the weight of the loaded trailer onto the rear wheels of the Ferguson tractor. In our second month of farming at Kensham Farm, in October 1956, I learned the lesson the hard way that matching implements to the tractor for any specific task was the only way that would be safe on the sloping fields of the Chiltern Hills.

It was when I was drawing the four wheeled trailer loaded with bags of seed corn down one of our fields early on a frosty morning that October that the heavy four wheeled trailer took charge of the much lighter tractor – so that I knew for a few terrible seconds that there was going to be a most awful accident which it was too late to prevent. The trailer caused the tractor to roll over leaving all four wheels in the air. I was very fortunate to be able to walk home.

I explained to Alison that there had been a bit of an accident with the tractor, but then found that my appetite for the breakfast to which I had been looking forward had gone, and that my leg had been so badly bruised that although I walked home on it, I could not walk on it for the next week or 10 days. I can remember Alison walking with me along the main road to the nearest telephone box, myself on crutches and Alison helping me along, so that I could make our routine weekly telephone call to my parents to explain that everything on the farm was going splendidly - with no more than a passing reference to a bit of trouble with the tractor. Those were the days when it was difficult to get a landline for a phone, and mobile phones had not been invented. Soon afterwards our neighbours at Hill Farm, Cadmore End, Sidney and Monica Lacey, agreed to share their telephone line with us, on an arrangement known as a 'party line' so that only one of the two families could use the phone at the same time. Our predecessors, the Bird family, never had any form of telephone at all.

Before the end of the 1950s, it became quite clear to me that there was no longer a case for traditional mixed farming. I realised that we had to specialise on one core enterprise which was to be the production of milk, and that it would be best to have no more than one supplementary enterprise. We decided that either the laying hens or the pigs had to go, to be followed by increasing the scale of one or the other enterprise.

It was then that I had the fortunate opportunity to buy four ex-military huts, each of them 20 feet wide by 60 feet long in sections for which I paid £120 for the lot. There was a cost involved with a local builder to erect 4-inch concrete block dwarf walls on footings, but without any concrete floors, and for the sections of the four huts to be erected on the dwarf walls and then for the hut roofs to be waterproofed with new roofing felt. Nevertheless, our four first deep litter houses, each sufficient in size for around 500 laying hens, enabled us to start serious egg production with minimal costs.

The pigs had to go. I raised some cash by selling the five pig farrowing huts to a farmer near Checkendon in Oxfordshire and we increased the numbers of dairy cattle producing milk for the Milk Marketing Board - at that stage the milk from us and neighbouring farms was allocated to Jobs Dairy in Middlesex. We sold our eggs wholesale to a producer cooperative, Thames Valley Egg Producers Limited, near Didcot. I remember visiting the headquarters and packing station and hearing from the founder and managing director, Gus Kingham, how he had formed the Cooperative in 1934 to market the eggs produced by the farmer members effectively, since before then farmers in any given locality were all cutting the price of their eggs to compete with the neighbouring farmer's price for sales to local shops and households, until there was no profit margin in producing the eggs for any of them.

Section 2, 1960 to 1969 – Integrating Watercroft Farm with Kensham Farm

In 1960 the farmer who owned the adjoining Watercroft Farm of about 104 acres came to me one Friday evening to explain that his wife was not happy at the farm and he intended to move and to put the farm on the open market after the weekend through agents.

I wasted no time in showing my interest in this land and in consulting with my father to ascertain whether we could afford to buy it for the asking price of about £9,000 for the freehold. My father found that the best way of buying this adjoining farm was for my grandfather to raise the money from other investments. In that way Watercroft Farm became part of his estate, and thereby reduced his potential liability for estate duty. By then my grandfather, Walter Edgley, was already just over 90 years old - he died later in that same year, 1960, having known that he had helped his grandson's new enterprise in a tax efficient manner.

In the farming press in recent years many articles have appeared extolling the virtues of succession planning. For our family, this example of succession planning sixty years ago by my grandfather Walter Edgley was a precedent which we in our family have always tried to follow.

The purchase of this additional land proceeded with speed and without Watercroft Farm ever reaching the open market. We then had to integrate the two farms into one single farm unit of 206 acres in total, rather than the land remaining as two adjoining farms of just over 100 acres each.

We facilitated this integration by realigning various field boundaries at Watercroft Farm and setting up an entirely new straight track along the ridge of the land starting at Kensham Farm and passing right along the ridge of Watercroft Farm, ending at the top of a Forestry Commission wood known as Barn Wood. In those days, grubbing up hedges was encouraged so as to increase the productivity of the farms, and so we received a Ministry of Agriculture grant towards the cost of grubbing up the crooked hedges, thus enabling us to make the new track with nothing more than some hardcore to firm up the softest places.

With the new field alignment, we devoted the Kensham Farm acres to the dairy herd for grass leys of two to four years duration for grazing and for cutting for silage, with one field each year for kale, which the cows grazed with a fresh strip each day in front of an electric fence that the herdsman had to move each morning during the winter months. On the adjoining Watercroft Farm we grew a rotation of cereal crops, the 4-year rotation being wheat, followed by a second crop of wheat, then barley, and finally oats before returning to wheat for the start of the next four year cycle of the rotation.

In those days tractors were small, field sizes were small, and it was still the norm to farm with an adequate staff, for which wages were not the significant item that they became in later years as a charge on farm accounts. For our first harvest at Watercroft Farm we bought our own combine harvester - it was a second-hand PTO driven, 5-foot cut, Activ combine, to be drawn and powered by our elderly Nuffield tractor.

For our increasing production in the 1960s we employed more staff, most notably in February 1962 when we employed Nigel Rogers as a young tractor driver who would live in one of the pair of new cottages that we had built at Watercroft Farm. It had been simple to get planning permission, since at that time local authorities seldom refused an application to build cottages for farm staff. Nigel was to remain as our foreman for the whole of his long working life. He drove our first combine harvester in 1963, in the years when I drove the tractor with grain trailer and ran the grain dryer. Nigel continued to be our principle combine driver until the harvest of 2018, before handing over those duties to his son, Paul Rogers, who had already taken over as our farms' foreman several years earlier.

It was in 1963 that we took the decision that it would be best to separate our farms at Cadmore End from The Sonning Land Company Limited, and so we set up a separate Limited Company in which Alison and I were majority shareholders and my parents were minority shareholders. My father carried out all the legal work from his office in London and it was interesting that our accountant advised that it would be simpler to buy the shell of a defunct company, rather than set up a new company from scratch. We therefore bought the shell of 'Moir Haskew Farms Limited', a company that had farmed unsuccessfully at Sydenham Grange near Thame in Oxfordshire, for £100. We subsequently had a stroke of good fortune when we found that the accumulated tax losses of the company's earlier trading as Moir Haskew Farms Limited could be used to offset some of our profit in subsequent years trading as Kensham Farms Limited.

In 1968 we employed our first farm secretary, Mrs. Whittles, before the days of electronic adding machines and computers. Mrs. Whittles used a comptometer, an early form of calculator that operated without any electricity on a system of mechanical ratchets. With the aid of this machine, which Mrs. Whittles operated with the speed and power of a machine gun, she was able to carry out all the calculations for keeping our farm accounts in cash analysis books. Mrs. Whittles did not retire until she was into her 80s at which time she was succeeded by Mrs. Chris Wheeler who has embraced new computer accounting systems with the specialist software firm Sum-It so that no cash analysis book has been used in the farm office since 1993. During the years 1955 to 1968 I carried out all our farm recording myself, and I do realise how fortunate I have been to have had only two farm secretaries to keep all our farm records and costings in the 53 years since then.

The 1960s were the years when we greatly expanded our egg production activities. The Hatchery producing day-old chicks for use within the poultry farming sector, Thornbers, based at Mytholmroyd in Yorkshire, was expanding its sales of hybrid egg layers. One of these was the 'Thornber 101', a small hybrid bird based on White Leghorn blood which was selling well since it was able to produce more eggs than conventional pure-bred pullets. Thornbers were expanding in those days, and had built a new hatchery at Charvil, near Sonning in Berkshire. Interestingly, the Charvil Farmhouse and small steading which Thornbers bought as the site for its new hatchery had once been the farmstead for a large farm with around 600 acres bounded in the North by the River Thames which my grandfather, with a wealthy investor, had bought to form the original Sonning Land Company Limited in the early 1920s. When that company came close to failing in difficult times in the 1930s it was only my father's administrative ability, rebuilding the company over a period of some thirty years with some modest residential properties in outer London, that enabled it to survive and eventually succeed.

When Thornbers set up this new Berkshire hatchery they looked for suitable 'Supply Farms', that is independent farmers willing to rear and mate the breading stock for its hybrid birds on a deep litter system to supply the new hatchery with fertile eggs. We signed up as one of the first of these Supply Farms. The essence of Hybrid breeding is to use a cross between two different lines to produce the hybrid with reliable characteristics. It is not possible to cross the progeny of first generation with each other to breed the next generation with the same known productive characteristics. It is always essential to go back to the lines of original parent stock to produce the hybrid as a first cross. For this reason, Thornbers supplied us with day-old chicks from the female line as well as from the male line which we had to rear in separate pens in the brooder house and then in outdoor pens to around 18 weeks old. At that stage, just before point-of-lay, we put 25 of the young cockerels into a deep litter pen with 250 of the pullets, this being the standard ratio in poultry breeding of one cockerel to look after ten pullets to produce fertile eggs. These fertile eggs would be collected by Thornbers for hatching out ready for sale as Thonber 101 day-old chicks to the farms that produced eggs for human consumption.

After a few years, when Thornbers faced severe competition from other hatcheries producing hybrid day-old chicks, they reduced the number of hatching eggs that they required from us. We had already built our first battery hen house in a building 110 feet long by 20 feet wide, containing two banks of 4 tier cages to house around 3,000 pullets for commercial egg production, and had cooperated further with Thornbers by using our battery house as a 'Test Unit' for Thornber breeding stock. For this we had to fill in a record card for eggs collected from each cage separately, to provide Thornbers with data on the merits of 20 different strains of birds.

In 1961 I was elected as Chairman of our local High Wycombe Branch of the National Farmers Union to serve my first stint of two years. It is interesting to reflect that at that time most farms were small, the average farm being between 100 acres and 200 acres, so in any given area there were far more farmers than nowadays. I was re-elected to serve a second stint of two years precisely 50 years later, in 2011, but by that time, following the many farm amalgamations that had taken place, the local branch covered the whole of South Buckinghamshire and Middlesex rather than just High Wycombe. I was delighted when my oldest grandson, Alex Nelms, who now works with us in the Kensham Farms partnership and is taking over the management of our diversified farm enterprises and land agency matters from me, was elected to the same office as NFU Chairman of our South Bucks and Middlesex branch in 2019.

By the end of the 1960s I made more of my own time available for off farm activities, including service as a Governor of Cadmore End Church of England School, and as a part time Tutor at HM Borstal, Finnamore Wood, where I taught mathematics, photography, tractor driving and also led group discussion in small groups of 6 to 10 trainees, all of whom were within the age bracket of 17 to 22 years old. Some of our lads regarded that penal establishment as if it were going to college and many of them, after discharge, met the right girl and made sense of their future lives. I was to go on teaching on a part time basis at that penal establishment for a total of 22 years, during which time it ceased to be called 'borstal' and was renamed as a 'Young Offenders Institution'.

On the family front in the 1960s, Rosalind and Paul started at the Cadmore End Church of England School and we were delighted when Alison became pregnant for a third time and our son Charlie was born in 1963.

I will explain in later sections of this chapter how Charlie joined our farming partnership on completing the three-year course at the Royal Agricultural College in 1985 and took over the management of our real farming enterprise, the arable crops, many years ago.

Having described how we established our farming at Kensham Farm I will glance over the remainder of the 65 years in which Alison and I farmed the land and brought up our family, with a few notes of the main happenings of each subsequent decade.

Section 3, 1970 to 1979 – Tenancy of adjoining land at Dells and Bigmore Farms taken on. Sale of Kensham dairy herd and the change to arable farming

In March 1970 we had the opportunity to rent Dells and Bigmore Farms which shared common boundaries with Kensham Farm and was only separated from Watercroft Farm by a narrow shaw of Forestry Commission woodland, through which we were subsequently able to negotiate a formal Easement for Right of Way from the Forestry Commission.

This new land became available when my friend, John Rowntree, moved to Sussex following the death of his landlord Sir Francis Leyland-Barrett Bart. John was able to buy the farms as sitting tenant from the Executors of Sir Francis and then took the decision to place them on the open market with vacant possession. We were not in a position to be able to make an offer for these adjoining farms at that time, but in 1973 they were bought by an industrialist, Brian Ball-Greene, who wished to hold the land as an investment and to be able to ride his horse around it without farming the land. We already knew the Ball-Greene family and we were pleased to be offered a tenancy of the land, an expansion which would increase our overall farm acreage up to 446 acres. The rent for this additional 240 acres was £1,700 per annum, this equating to just over £7 per acre per annum.

The farming of the additional arable land at Dells and Bigmore Farms changed the entire emphasis of the farm from dairy farming to growing of arable crops. The only way we could raise the working capital necessary for the first year's cropping was to sell the dairy herd, since with arable farming all the expenses for the year have to be paid before there is any harvested crop ready for sale. Additional investment was required for farm machinery, seed, fertiliser and crop protection chemicals, as well as converting the cattle yard into grain stores with wood end barriers and a galvanized sheet iron tunnel with 'A' shaped laterals across the floor to dry the grain.

Messrs. Thimbleby & Shorland of Reading held a dispersal sale of our herd on Wednesday 21st January 1970 here at Kensham Farm to which buyers travelled from as far as Shropshire and Wales. Our 111 accredited Friesian cattle, including 41 cows or in-calf heifers and 40 unserved heifers, yearlings and weaned heifer calves fetched good prices at that time. I still have the sale catalogue marked up by the auctioneer, Kerr Kirkwood, showing that the sale total was £14,024-11-9 (in pounds, shillings and pence, the sterling currency that was still in use at that time) with the 70 cows averaging £131-15-9. We also sold 56 store cattle, the dairy plant and our Galloway bull *Highwayman of Ardoch*. I still remember the occasion in 1966, when my daughter Rosalind, who was 10 years old, and due to have her piano exam later in the same day, helped me to herd *Highwayman* back to his correct field after he had broken out. I regret I do not recall what grade my daughter achieved in her piano exam.

In 1975, we were able to buy the Manor of Cadmore End which is Common Land to the east of Kensham and Watercroft Farms and provided an access track between the farms. Our low-key management of this Common has included restoration of three ponds and management of the trees on the Common, partly funded by a Forestry Commission England Woodland Grant Scheme. The footpaths and the whole Common is appreciated by local dogwalkers, and we have made permissive bridleways for horse riders so that the footpaths are not churned up by horses' hooves.

In 1987 Alison and I set up the Little Common, Cadmore End Residents Association, so that the Association could take on the responsibility of repairing the track which gives access to many of the houses in the village of Cadmore End. We hand delivered a full letter of explanation about the track maintenance, together with a calling notice for a meeting to be held in the Cadmore End Church Hall. We were particularly pleased with the response when most of the houses in the village were represented, and at this first meeting a Chairman, Treasurer and Secretary were all appointed, together with a few further members to serve on a committee. This structure has worked well ever since, with an AGM each year at which a membership fee, payable by all homeowners in the village, is agreed to cover track maintenance charges for the ensuing year. The Association also publishes a village newsletter, and in recent years has organised events in the Church Hall such as local history displays and lectures, or on Little Common for summer barbecues.

Section 4, 1980 to 1989 – Charlie Edgley becomes a Partner in the Kensham Farms Partnership. Purchase of the freehold of Dells and Bigmore Farms, and Penley Hollies land

Now that arable farming, the growing of crops mainly wheat but also with some barley, oats and oilseed rape, had become the main focus of the farms we no longer required staff in all our farm cottages. In the early 1980s there were great difficulties over letting a spare cottage, in that once a tenant was offered a tenancy of a house or cottage, the tenant could become a tenant for life, with the landlord unable to regain possession should he require the cottage for other uses in the future. We wanted to draw some income from the spare cottages without the threat of losing possession, and so we let rooms in the spare cottages as students' accommodation under Licenses to Occupy. When the Housing Act 1988 was introduced subsequently, it became possible to let a spare house or cottage on Assured Shorthold Tenancy (AST) terms and for the landlord to regain possession under Section 21 of the Housing Act 1988. Regrettably the present government has made proposals to remove the Section 21 provisions, apparently having forgotten that it was this provision of the 1988 Act which made so many houses and cottages available for tenants to rent.

It was in April 1980 that we were able to buy the Freehold of Dells and Bigmore land that we had been renting for 10 years from Brian Ball-Greene following his decision to move to the Hambleden Valley. For this purchase we took out an Agricultural Mortgage Company (AMC) mortgage at a time when runaway inflation was occurring so that the first half years interest rate was just over 18% per annum. In 1982, we bought a further 12 acres of land that at that time had been called Oakfair Farm which adjoined Dells Farm. In January 1983, we started a system of Quarterly Reviews with which we could check our costings for the different farm enterprises each quarter.

In 1985, our son Charlie completed the three year sandwich course for the Higher National Diploma (HND) in Agriculture at the Royal Agricultural College. During the middle practical year at that course, Charlie had worked on the sheep station owned by John and Joanna Williamson in Australia. Joanna's grandparents had met my grandparents when they visited Australia in the 1930s and both families have kept in touch with each other ever since. In the same year we bought a further 73 acres of land, known as Penley Hollies, from the sale of Sir David Brown's Estate at Chequers Manor, Cadmore End, for £1,706 per acre.

My father, Roy Walter Kelsey Edgley, died on 9th February 1986. It was his work that had enabled me, with my Alison, to set up these farms in 1955 with initial finance provided by him and his company, The Sonning Land Company Limited, for the early critical years before the farms were paying their own way. My father retained his interest in the work of our farms throughout his life, and was still attending to family legal matters a few days before he died.

It was during the end of the 1980s that we started to introduce diversified enterprise when we found that we had spare buildings that could be used as craft or light industrial workshops not connected with the farm. It was also in the 1980s that we received a demand from the Diocesan Church Authority for us to repair the roof of the chancel of St. Margaret's Church, Lewknor, around 9 miles from Kensham Farm in the next county of Oxfordshire. I was most aggrieved at that demand, since the deeds of our land showed no mention of any chancel repair liability. I was all the more aggrieved since by then I was a Churchwarden of our local parish church, Holy Trinity at Lane End, and also served the Diocese in other ways as a Governor of Cadmore End school, as photographic editor of the Oxford Diocesan magazine and as a member of the Diocesan Parsonages Board. In the next chapter of this book, I will describe the way in which I attended to this demand, with the ultimate result that the laws of England have been revised – albeit that it took 30 years to achieve that favourable result.

Section 5, 1990 to 1999 – Cessation of egg production and purchase of Top Plain land

We were able to increase the area of arable that we farmed in the early 1990s by renting fields near Harecramp from two neighbouring landowners on *Gladstone v Bower* terms.

This type of tenancy made use of an apparent flaw in the provisions of the Agricultural Holdings Act 1948, by which it became possible for a willing landlord to let to a willing tenant an agricultural field or holding for a term certain of more than one year, but less than two years, without the tenant having the legal right to turn the tenancy into a year-on-year tenancy.

The Agricultural Holdings Act 1948 had almost certainly intended that all tenant farmers, occupying land under an AHA 1948 tenancy, would have security of continuing tenure, whether or not the landlord might wish to end the tenancy. At that time landlords were reluctant to grant any new AHA 1948 tenancies, since they might never again be able to occupy their own land or use it for any other purpose. That legislation had the adverse effect of no land becoming available for aspiring young farmers to start farming as a farm tenant.

Gladstone v Bower was a 1959 case in the English Court of Appeal which established that, although AHA 1948 had made provision for farm tenancies of less than one year, and also tenancies of over two years, to automatically become year-on-year, strangely those provisions did not apply to tenancies for a fixed term of more than one year but less than two.

The Agricultural Tenancies Act 1995 superseded the earlier Acts, and new tenancies, which became known as Farm Business Tenancies (FBTs) have been welcomed into general use and have had the effect of more farm land being available for potential farm tenants. This has been the result of the landlord being able to agree with the tenant the length of tenancy, in the knowledge that Government will not then override the length of tenancy set out in an FBT agreement.

It was in 1993 that we took the decision to cease egg production. The three main reasons for this decision were firstly that my son Charlie, who would clearly be the future of the farms, disliked chickens! Another important reason was that the smaller greengrocer and butcher's shops trading locally from whom we had a weekly order of two or three cases of sixty dozen in each case, were increasingly going out of business following the new method of retailing food in supermarkets. Our head poultry man, Michael Leaver, with his wife took over the sale side of our egg production business and traded in their own name for some years, before Michael finally came back to Kensham Farms to run our grain dryers in a part time capacity.

It was in 1996 that we purchased a further 123 acres of Sir David Brown's former estate, Top Plain, from a golf course development company. This land once had grass gallops around its perimeter on which one of Sir David Brown's horses had been trained prior to winning the Cheltenham Gold Cup at the end of the 1950s.

In 1996 we built our Lingward grainstore with underfloor ventilation system and tunnels, enabling us to dry the grain in a similar way to that on which housewives' clothes lines used to work – that if there is a good breeze of dry air then drying gradually takes place.

Section 6, 2000 to 2009 – First harvest from land rented from the West Wycombe Estate

It was in the Autumn of 1999 that Sir Edward Dashwood Bt. advertised in the *Farmers Weekly* for a tenant to take on the farming of his land on the West Wycombe Estate. We were keen to tender for this opportunity to rent a significant additional area of arable land, since the western part of the West Wycombe Estate adjoins the northern part of our Watercroft Farm. We were told that there had been 75 enquiries to the advertisement, with a dozen serious tenders. We were immensely pleased when Sir Edward offered my son Charlie a Farm Business Tenancy (FBT) of 1,300 acres of the arable land of the West Wycombe Estate for us to farm as part of our Kensham Farms holding.

It was in 1999 that we asked Paul Rogers, son of our Foreman Nigel Rogers who first joined us in 1963, to come back from his civil engineering job to work as tractor driver on the new greater acreage that we were now farming. In subsequent years, Paul has taken over from his father Nigel as Farm Foreman, but he did not take over driving the combine harvester until the harvest of 2019. Before then Nigel had combined every single harvest for us between 1962 and 2018, the first with a small Active five-foot cut combine harvester and the latest with a John Deere S690i with a thirty-foot cut header. The capacity of the present John Deere combine harvester is such that in ideal conditions it can harvest one tonne of grain every minute, but its normal output is between forty-five and fifty tonnes harvested per hour. This means also that we have had to improve our trailers, which now have a capacity of sixteen tonnes rather than our earliest trailers that held three tonnes, and we have had to increase grain handling mechanisms at the graindryer and store to be able to receive grain at the same speed at which the combine harvester harvests it.

We also took on a further Farm Business Tenancy of land at Chawley Manor Farm which adjoined the West Wycombe Estate land. For this additional cereal acreage, which brought our total arable acreage up to the order of 2,000 acres, we had to make the decision whether to use two moderately big combine harvesters or one really big combine. We have always been pleased that we took the latter decision, initially with a Deutz Fahr combine, then with Claas Lexion combines, but more recently with John Deere combines which we tend to keep for three to five years each before part exchanging for a new model.

It was in 2005 that we started to use Global Positioning Satellites (GPS), the very same technology that is used for a sat nav in a motor car, for our arable farming tractors and combine harvesters. We now have a satellite receiver on the roof and computer inside every tractor cab. This GPS technology is quite remarkably effective in that if we go into a field that we have not previously cultivated and drive once around the perimeter of the field, the computer will then make an accurate plan of the field. When it comes to operations such as seed drilling, where accuracy is of the utmost importance to prevent any overlapping or missed patches, the computer can be set so as to show the position of the tractor on the screen with the planted part of the field in one colour and the part yet to be planted in another colour. The computer technology will then steer the tractor on all the long runs so as to be the precise distance from the last bout in order to achieve the aim of no patches double drilled and no patches missed out. At the end of the runs, the tractor driver takes over the steering wheel to manually steer the tractor into the correct position for the next long run.

Section 7, 2010 to 2019 – Purchase of land at Hill Farm and Building New Grain Stores at Kensham Farm

It was in January 2011 that I had the honour of being invested with Membership of the Order of the British Empire (MBE) by Her Majesty the Queen at Windsor Castle 'For services to the agricultural industry and to the community in Buckinghamshire'. On the day of the investiture I was allowed to be accompanied by just three guests. This could have been difficult within the family but my son Paul had teaching commitments at Christ College Brecon in South Wales which he would have found difficult to miss and so I was accompanied by my Alison with our daughter Polly and my son Charlie who drove us to Windsor Castle in our old Volvo estate car. After the investiture we were joined by other members of our family for lunch at the Phyllis Court Club at Henley-on-Thames.

In 2017 a further neighbouring farm, Hill Farm at Cadmore End was on the market. This was the farm which used to be farmed by Sid Lacey and his wife Monica, the couple who had been kind enough to share their telephone line on a party line arrangement with us when we first came to Kensham Farm in 1955. I am still in touch with their son David Lacey who briefly worked for us in 1955, prior to emigrating to New Zealand on a £10 Government assisted emigration package that was widely used by English folk wishing to emigrate at that time. I could add as an aside that his mother took him and his small suitcase to the bus stop at the Peacock Inn at Cadmore End where the bus conductor (in those days a bus had a driver and a conductor) asked him if he wished to go to High Wycombe – to which David replied: "No, I wish to go to New Zealand!" David was fortunate in meeting his wife Phil relatively soon after arriving in New Zealand and together they established a successful farm and reared a large family, and we have remained in touch ever since.

It was for Autumn 2019 that we bought a Dale direct drill which had a knife coulter rather than discs to cut the narrow channel into which the seeds are blown pneumatically. We have found this to be a great advantage with autumn drilling when conditions are damp and the seedbed in less-than-ideal condition. With a really dry friable seedbed the discs on a drill are able to make an excellent and even job of placing the seeds, but often we get autumn conditions in which the soil is slightly sticky rather than quite dry and in these more difficult conditions the seed drill with a knife coulter can keep going satisfactorily without the blockages which are apt to occur with disc coulters. This may seem a small point, but it makes a huge difference in that with an autumn in which the soil remains sticky the seeds can be drilled, whereas farmers using the disc coulter and waiting for ideal conditions may often find that the autumn planting has not been achieved, leaving fields which then have to be seeded the following spring.

In 2013 we greatly improved our grain dryers and stores with a new grainstore in our small field next to the M40 motorway, this store having a capacity of three thousand tonnes, equipped with a Scandinavian Svegma grain dryer which has a throughput of forty-six tonnes per hour. The Svegma dryer, as a continuous dryer, works with hot air and completes the drying process as the grain passes through the dryer in an hour or less, whereas with our earlier grainstores with underfloor ventilated drying, moisture extraction was a slow process sometimes taking two or three weeks.

The new grainstore and Svegma dryer has been so successful that we found we could build an additional store for 1,200 tonnes of grain in June 2019. This adjoins the 2013 store, with the grain being dried in the same graindryer from which it is then moved to the new JRC store using the forklift with grain bucket which has a capacity of 2.4 tonnes in each scoop.

The size of grain lorries has increased over the years so that the present normally used grain lorries can carry a load of thirty tonnes of grain, which with the weight of the lorry, means a total of over forty tonnes. Some farm driveways are badly placed for traffic of this size, but we have the advantage of a yard at Kensham Farm in which we can handle these large lorries which then travel a distance of only some five miles along the B482 road to Stokenchurch where they are able to get onto the M40 and the system of motorways now serving the country.

I must end this description of the history of our own farming at Kensham Farm with the very satisfactory family news that my grandson Alex Nelms joined us in the farming partnership in 2013 following his degree course at Aberystwyth University and a further estate management course at Harper Adams University, and then Charlie's son Angus joined us in the farm partnership in 2019 after completing the three-year sandwich course in Agriculture at Plumpton College in Sussex.

Kensham Farmstead in 1956, following clearance of some overgrown hedges around the former stackyard.

Rebuilding the unsafe southeast gable of the farmhouse in 1956. Acrow props were used to hold up the cruck beam and subsequent 16th century massive beam which forms the ceiling of the present dining room.

Dell Brothers of Lacey Green carried out all the initial renovations of the farmhouse and buildings within the farmstead. When the cattle manure had been cleared from the yard, a concrete apron was laid next to the buildings, and hardcore topped with gravel was laid in the yard. The Ford van in the photo was used by Bryan and Alison for a trip to Scotland for their honeymoon, and was the only motor vehicle for their first two years at Kensham Farm.

Landrace in-pig gilts and the sectional pig huts built by Bryan in 1955. During the late 1950s the decision was taken to specialise on the production of milk and eggs and the pigs and pig huts were sold.

The first Crendon Concrete building, built in 1956, to house the milking parlour and dairy, with external stand on which the full churns of milk were put out each morning ready for collection by the Milk Marketing Board's haulage contractor.

Bryan with Alison in the early days of taking over Kensham Farm, holding one of the breeding stock pullets for production of Thornber 101 hybrid laying hens. The pullets were reared firstly in Tier Brooders kept at a warm temperature indoors, and then transferred to Haybox Brooders and outdoor rearing pens (where this photo was taken) until point of lay. At that stage they were moved indoors to deep litter houses together with the cockerels which had been reared separately, for the production of fertile eggs.

Photo taken by Alison of Bryan with their daughter Rosalind (Polly) on holiday in 1964.

Charlie Edgley, about to take a stem of marrow stem kale to show to his teacher and friends at Cadmore End Church of England School in 1968. The builder's dump truck, the back of which can be seen at the right of the photo, was modified in the farm workshop to carry 1 tonne loads of layers' mash, milled in the farm Old Granary, round to the layers' poultry sheds.

Harlow Brooder at Watercroft Farm in the late 1960s. The tube feeders had been put outside while the deep litter was being removed between intakes. Three deliveries of 3,500 day-old chicks were taken in each year for rearing up to 15 weeks old, when they were taken in crates to the battery cages at Kensham Farm for egg production.

One of the sections in the Andover Deep Litter building housing 220 of the Thornber 101 female breeding stock pullets and 22 Thornber 101 unrelated breeding stock cockerels. The fertile eggs produced were then collected by Thornbers, and would be incubated at its Charvil, Berkshire Hatchery. The day-old Thornber 101 hybrid pullet chicks would then be sold in boxes for delivery to commercial egg producers.

Table insert text:

BATCH | KENSHAM FARMS LTD. | WEEK ENDING
BREED | NOTES | TOTAL EGGS
HOUSE | ON a.m./p.m. OFF a.m./p.m. LIGHTS | PULLETS END OF WK.
PULLETS START OF WK | MIN. TEMP MAX. AGE. END WK. | % PROD. HEN DAY

One of the banks of 4 tier hen battery cages for commercial egg production. The brown colour birds in this 4 tier stack of battery cages were Thornber 404 breed, laying brown eggs, whereas the Thornber 101 hybrid, based on white leghorn lines, laid white eggs. We always recorded all the important production factors of every flock on weekly record cards to our own design (shown as an insert). Before the advent of electronic calculators, I used a slide rule as the simplest method of calculating weekly percentage production of each flock, with the results plotted on a large graph, with each flock colour coded.

The Ben Nevis egg grading machine in the Egg Shop sorted the eggs from a conveyor belt into the different weight categories of small, medium, large and extra large. The standard size egg trays shown in the photo hold 30 eggs.

Rearing day-old chicks under Calor Gas brooders in the Harlow Brooder shed at Watercroft Farm. The heat under the brooders was designed to replicate the warmth under a mother hen's wing. The hardboard surrounds were removed after two weeks, so that the growing chicks could move about anywhere in the brooder shed.

Baby chicks the day after delivery from the hatchery. The lids of the boxes in which the chicks were delivered were used as feed trays, until the chicks had grown sufficiently to use tubular hanging feed hoppers. One of the black suspended water troughs can be seen on the left.

A former Brooke Bond tea delivery Trojan van was converted to take the layers mash from the granary to the six different hen battery houses.

Bryan and Alison, with head poultryman Michael Leaver and his son Mark in the farm egg shop. This photograph was published in the Bucks Free Press in December 1988, at the time of the Salmonella in eggs scare falsely raised in the House of Commons by Edwina Currie MP. In the accompanying news article by Eveleen Houdret she pointed out that the scare had arisen from faulty hygiene in the House of Lords kitchen, which caused 20 peers to be infected by the bug Salmonella Enteritis. Eveleen noted that the chance of eating an infected egg was 1 in 200 million, and that sales of our Kensham Farms eggs to over 20 local greengrocers and butchers had not been affected.

Milling and mixing poultry mash in the Kensham Farm Old Granary. The 15hp electric hammermill blew the ground wheat or barley after milling straight into the 1 tonne food mixer shown into the photo. The hanging dust bags allowed the air to escape while trapping any flour still in the air. Our normal weekly output was 12 mixes of 1 tonne each.

When we bought our first fork lift truck, a Sambron, we made a platform for it of a size to take a 1 tonne load in open ended bags, all kept in place by the weld mesh siders, standing upright. At that stage both the Brooke Bond trojan van and the modified builders dump truck were scrapped.

The Kensham herd of Friesians grazing grass in early Summer. When these fields were used for arable crops, following the sale of the herd in 1970, the fences were removed to make 2 fields of just over 20 acres each rather than 5 smaller fields.

Photo of our Kensham herd grazing the fields now known as Landcraft in the early 1960s. The furthest away cattle are grazing the part of this field through which the M40 Motorway was built in the late 1960s. St. Mary-le-Moor Church and the village of Cadmore End can be seen in the background.

Our Friesians strip grazing a field of a cocksfoot lucerne mix behind an electric fence at Kensham Farm. At that time Dells Farm was farmed by my late friend John Rowntree, whose dairy herd of Ayrshires can be seen grazing at Dells Farm in the background.

During the late Autumn and Winter months, when grass does not grow, we strip grazed a field of marrow stem and thousand-head kale behind an electric fence. That kale was normally finished in late December or January. After that the cows self-fed silage from a large clamp of ensiled grass in the barn adjacent to the yard and parlour.

Silage made from grass, preserved in lactic acid which is formed in the fermentation stage within the silage clamp, is the main late winter feed for the dairy herd. The forage harvester which we used was a conversion of a David Brown Albion cutter. E & R Meakes Limited, our local blacksmiths at Lane End made an extended chassis trailer, with the flail cutter unit mounted at the front to convert this into the self-filling two-wheel trailer which I had designed. This design made it safe to use on the sloping fields of the Chiltern Hills, since half the weight of the load was transmitted onto the rear wheels of the tractor.

When the self-filling trailer was driven back to the clamp silo (in the barn on the right of this photo) it was tipped. A different tractor with buckrake was then used to move the cut grass up onto the silage clamp. We used to add molasses (black treacle) to assist the fermentation process that formed the lactic acid to preserve the grass as silage for winter cattle feed.

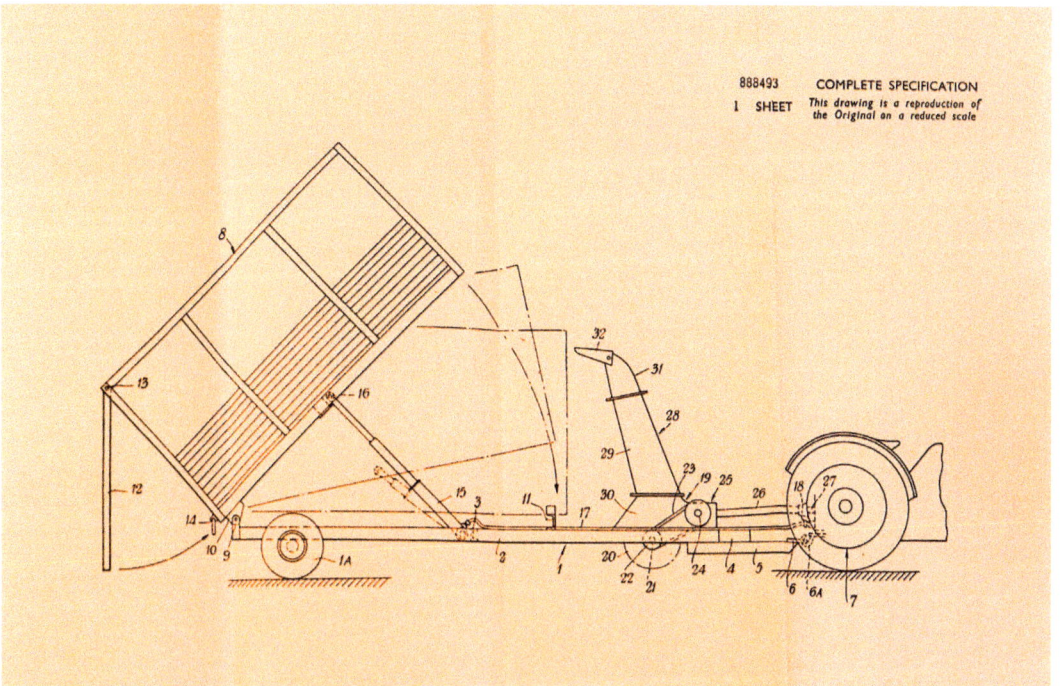

My father, Roy Edgley, arranged for a Patent Agent to take out a Patent after making a technical drawing of our blacksmith made self-filling silage trailer. We had hoped to interest a manufacturer in the design, but in this respect it was unsuccessful – possibly because on flat fields a tractor drawn cutter unit on its own wheels could then safely pull another trailer behind it, like a train pulling two carriages. However, we never regretted designing and making the self-filling trailer, which we used every year on our sloping fields until we sold the dairy herd in 1970.

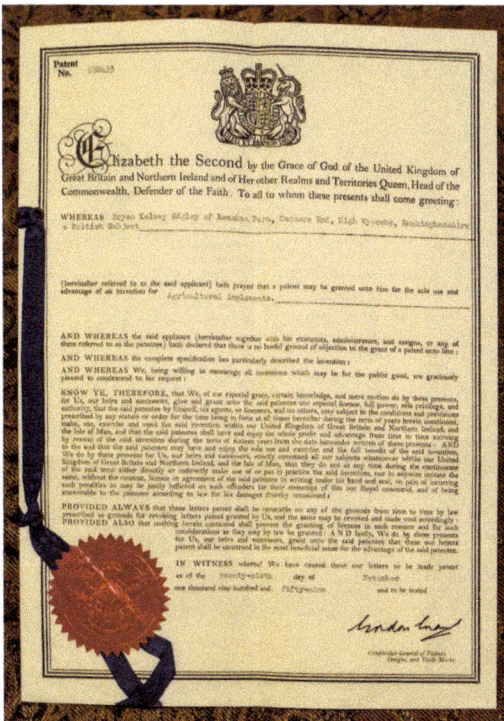

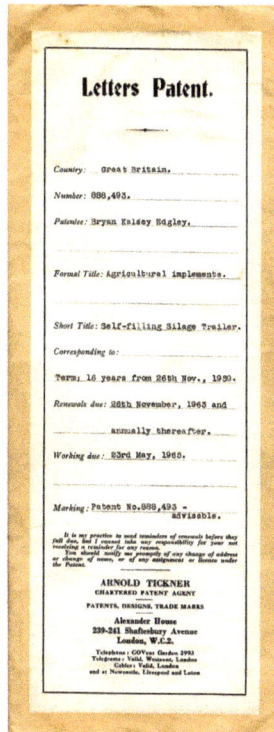

The Great Britain Patent No. 888,493 for the Self-filling Silage Trailer. The Patent was granted on 26th November 1959 but when it came up for renewal on 23rd May 1965, we allowed it to lapse.

77

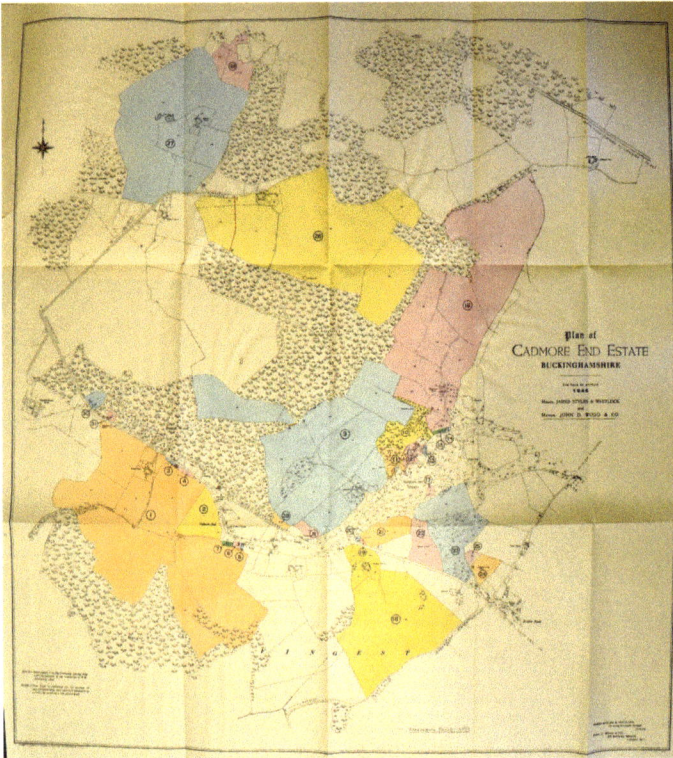

The Plan within the catalogue of the June 1945 Parmoor Estate sale. Kensham Farm which my father's company bought in 1946 is Lot 9 coloured in blue. Watercroft Farm, which we bought in 1960, lying north east of Kensham Farm was marked as Lot 16 coloured pink. It was owned by three different owners in the short time between the 1945 sale, and our purchase in 1960.

Dells Farm, lying north of Leygroves Wood and Kensham Farm, marked as Lot 26 coloured yellow, together with Bigmore Farm (Which is the unmarked land lying north of Leygroves Wood and Kensham Farm) was let to us in 1970 under a tenancy by Brian Ball-Greene, from whom we bought the freehold ten years later in 1980.

We bought our first self-propelled combine harvester, a 10ft cut Allis Chalmers in 1965, seen here at work at Dells Farm. We subsequently bought a Class combine harvester with 14ft cut and in the 1980s and 1990s used both combines for harvest.

Charlie, then aged 14 with his first car, an A30 which was no longer fit for use on the public highway, inspecting the crop of winter oats before our first harvest of the Bigmore Farm fields.

From the 1960s up to the end of the 1980s straw burning after harvest was considered to be good arable farm practice for controlling any residual weeds, thus making a clean stubble for the next crop. However, it had its drawbacks of removing organic matter which could have maintained humus level in the soil and it also emitted large amounts of toxic pollutants into the atmosphere. Furthermore, there was the danger of the controlled stubble burning fire getting out of control and causing damage to adjoining crops or property. Smoke emitted became a nuisance to private households nearby so Government bought in controls in 1991 to prohibit stubble burning at weekends or bank holidays, and two years later the Crop Residues (Burning) Regulation 1993 enacted a total prohibition of stubble burning in the UK.

Charlie offloading hay bales onto a Lister bale elevator with others stacking the bales in the barn in the 1980s. Materials handling has been improving over the years on most farms, so as to minimise the lifting of materials with human muscle power.

Our first Sambron rough terrain forklift truck being used to ease the task of loading the fertiliser spinner, by first lifting the whole pallet of fertiliser bags.

In the 1970s we used a Parmiter mobile grain elevator to load the grain stores from the trailers coming back from the harvest fields. The driver of the tractor with 3 tonne trailer had to tip the load quite slowly to prevent the Parmiter from being overloaded, and another worker had to level the grain out with a shovel in the grainstore.

The two new grainstores have concrete panel sides to a height of 4 metres. Removable timber barriers behind the roller shutter doors are fitted with hoops, used for manoeuvring them into position using the telehandler forklift. The trailers coming in from the harvest field, now with capacity of 14 or 16 tonnes, can tip straight into a large receiving pit below ground level. From this receiving pit the grain passes through a pre-cleaner (to blow out any chaff) and thence onto an overhead conveyor straight to the bunkers of 1,200 tonnes each. No manual lifting or levelling is needed in these modern grain handling systems.

The Watercroft Farmstead before renovation into craft workshops. The former Harlow Brooder, a corner of which can be seen on the right of the photograph was partitioned and windows were installed to form 4 of the 10 Workshops.

Farm Diversification with craft and light industrial workshops. This photograph shows the Watercroft Farmstead in which 10 workshops were set up in redundant farm buildings at the end of the 1980s. It will be seen that the former cowshed in the foreground on the left was re-roofed with a steeper pitch than the original, thus improving the visual impact of the building, increasing the internal height for use as a craft workshop, and providing a small high level window.

The craft workshops in use, Neil Harris, Picture Framer, assisting an artist customer to choose suitable pine moulding for a frame for her water colour painting.

Gary Kellynack and his father Alan, Furniture Manufacturers, making chair frames for a hotel chain.

Alison by the Land Rover with our then young Labrador, Gyp, on the new track at Watercroft Farm. This new track became immensely useful giving access to the different fields, after we integrated Watercroft Farm with Kensham Farm in the early 1960s.

We have approximately 11 miles of tracks around the farms for both access and horse riding. We sell either annual or single day Riding Permits to local horsekeepers. These riders are using the track described above, running along the ridge from Kensham Farm, through Watercroft Farm to Barn Wood.

At the time when we were re-roofing our Old Granary in 2000 as our Millennium Year Project, the pond in front of the farmhouse had become overgrown with reedmace. We enlisted the assistance of the Lane End Conservation Group to recommend suitable control. Tony Davis of LECG recommended spraying with Roundup Bioactive, but then there was a problem of how to apply this herbicide, since there was such dense vegetation growing out of the water.

The problem of how to spray the reedmace was solved by using a small dinghy, with one rope on its front at one side of the pond, and another at the stern to the other side of the pond. Tony stood or crouched in the dinghy with a knapsack sprayer, fitted with long reach lance.

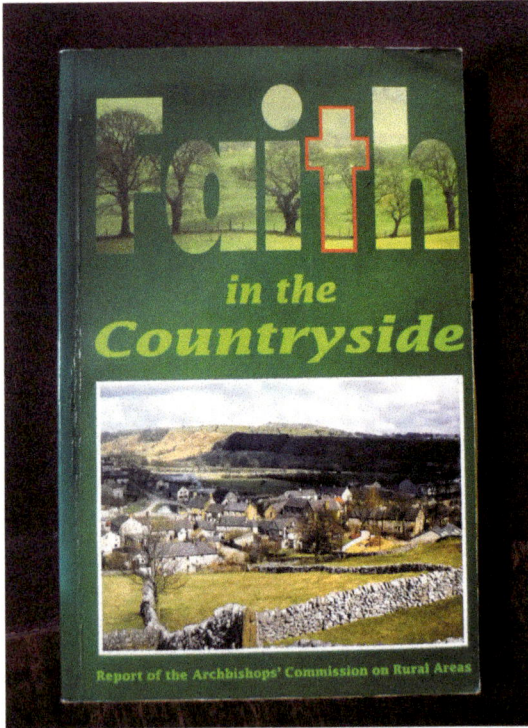

Faith in the Countryside, published in 1990 was a comprehensive report of some 400 pages on the structure and work of Anglican rural churches.

Bryan is now Churchwarden Emeritus of Holy Trinity Church at Lane End, having been one of the two active churchwardens for three years in the 1970s and for a second spell of 26 years ending in 2012. This photo was taken at the Remembrance Day service in November 2019.

The hutted Borstal Institution at Finnamore Wood where Bryan served as a part-time tutor for 22 years.

The parish church of St. Margaret Lewknor in Oxfordshire. It was a demand with threat of County Court proceedings against Bryan and others for re-roofing and repairs to make the chancel of this church wind and weatherproof that led to changed legislation in England concerning Chancel Repair Liability. Most unusually at Lewknor Church the chancel on the right side in this photo is taller than the main section of the church known as the nave.

Our project for the Millenium Year 2000 at Kensham Farm was to re-roof the Old Granary. I took this photo, of Alison and myself inspecting the work in progress, with the self-timer on the camera.

Our builders worked off a temporary floor, supported at eaves height by a 'bird cage' scaffold. Five new trusses to support the purlins and rafters, supporting Cellotex insulation boards beneath the battens. As many as possible of the original handmade clay tiles were reused.

Kensham Farmhouse. Alison in the garden with Councillor Darren Hayday and his wife Orsolya and their son who visited us to see the farm and farmhouse in 2014. Darren, who was Mayor of High Wycombe in 2006 is the great grandson of Jack Bird who, with his brother Frank, farmed Kensham Farm before us.

The cruck beam in our guest bedroom at Kensham Farm. The window just above the leg of the cruck beam is the window in the left gable of the farmhouse. This unusual construction is proof that the two gables of the front elevation cannot have been part of the original 'hall house' dating back to the 14th or 15th century. At that time the house would have been a simple hall, with the roof supported by this cruck beam which could originally have gone down to ground level. There would have been no first floor or chimney, but instead there would have been a hole in the roof to let out smoke from the open fire used for cooking as well as for warmth.

The M40 Motorway was built with only two lanes in each direction in the mid-1960s. By the 1980s it had to be widened to three lanes. This work made it necessary to demolish and rebuild several bridges. The Kensham Farm overbridge, which serves several private houses as well as the farm, had to be replaced with a temporary single lane sectional bridge while the work took place.

At night, between 10pm and 6am the next morning, the M40 had both lanes closed, while the original bridge, not much more than 20 years old, was demolished. All debris had to be cleared, and the motorway swept, before being re-opened for morning rush hour. Many Cadmore End folk viewed this work in progress at nighttime, standing on the temporary bridge.

Photo taken with the farm drone of the Kensham Farmstead in 2021. The two new buildings nearest to the M40 Motorway are grainstores providing storage for over 4,000 tonnes of grain.

Since the section of motorway crossing the fields farmed by us was curved, with much of it in a cutting, all the vehicles using the motorway lost their mobile phone signal, until two new aerials were built on our land at one end with a further radio aerial mast at the other end.

Charlie Edgley carrying 'A fair share of the bottle' placard in front of Big Ben at the NFU campaign on 15th March 2000 when the farmgate price for milk had plummeted.

Paul Rogers, son of our Kensham Farms Foreman, Nigel Rogers, driving one of our Massey Ferguson tractors drawing a trailer with large 'Fair share of the bottle' posters, around Parliament Square in London at the March 2000 protest.

Comparison of farmgate and retail prices 1990 to 2005

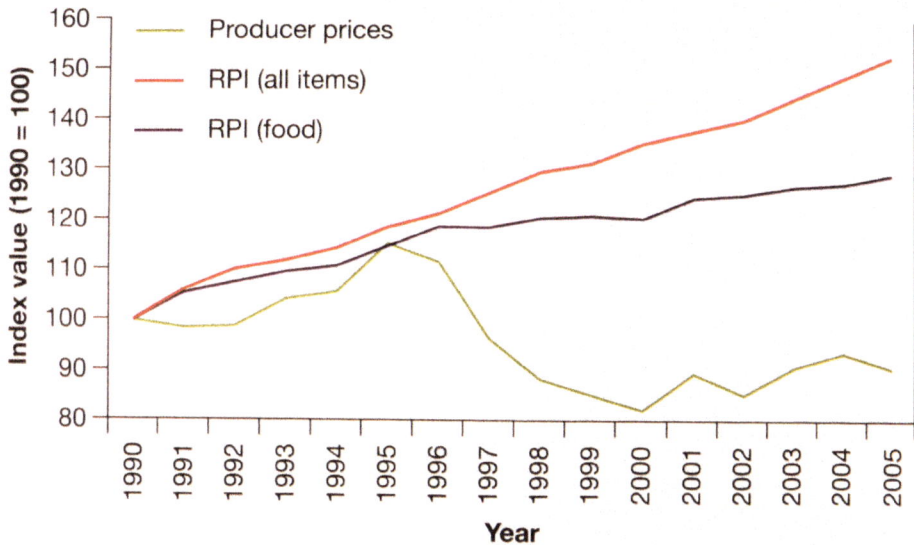

NB: RPI = Retail Price Index

This graph shows clearly how in the 15-year period starting from 1990 the Retail Price Index for food had risen to 133% of the 1990 price, whereas the return to the producer during the same 10 years had fallen to 82% of the 1990 returns. This had instigated the March 2000 NFU Fair Share of the Bottle protest and the Countryside Alliance Liberty and Livelihood march.

Our family took part in the Liberty and Livelihood march in London on Sunday 22nd September 2002. I took this photo of my family standing in front of the many placards which had done their job in Hyde Park. The Countryside Alliance had organised the protest, which was attended by 407,791 folk, largely from the rural areas. My family, from left to right are my daughter, Polly her three sons with one friend. Charlie standing behind my grandson William, with Alex on Charlie's right and my Alison on Charlie's left.

Bryan with Alison, Polly and Charlie, in May 2011 at Windsor Castle after Her Majesty the Queen had presented Bryan with the award of Membership of the British Empire (MBE) for services to the agricultural industry and to the Buckinghamshire community.

The Kensham Farms team at the Royal South Bucks Agricultural Association Ploughing Match and Show in October 2019 holding trophies awarded that year. From left to right Nick Perry, Alex Nelms, Paul Rogers, Charlie, Alison and Bryan Edgley.

Open Farm Sunday, 13th June 2010. Nigel Rogers, our farm foreman on the left with Steve Baker MP listening to the views of Cadmore End smallholder and gardener, Johnny Eldridge.

The French Ambassador, M. Jean-Pierre Jouyet, had requested NFU headquarters in May 2018 to arrange a meeting for him with a group of farmers. The Ambassador wanted to learn our views on the forthcoming 'Brexit' vote on whether the UK should withdraw from the European Union. We hosted the visit at Kensham Farm, starting with a full discussion in our dining room and ending with this photo: - Front row from the left, Alex, Jeff Powell (NFU County Chairman for Bucks, Berks & Oxon), the Ambassador, his official from the French Ministry of Agriculture, Charlie and Bryan.

Meeting Date:	4 July 2006	Open Gov. Status:	Open
Type of Paper:	Above the line	Paper File Ref:	
Exemptions:	None		

HEALTH AND SAFETY COMMISSION

A comparison of the risks from different materials containing asbestos

A Paper by: Kevin Walkin and Geoff Lloyd

Name of Board Member lead: Giles Denham

Cleared by Jonathan Rees on 21 June 2006

Issue

1. To provide further background information for the Commission to reach agreement on the risks from work with textured decorative coatings containing abestos (TCs); and to agree that a limit for sporadic and low intensity exposure should be included in the Regulations rather than the ACoP (in accordance with legal advice).

This is the front page of the Health & Safety Commission paper dated 4th July 2006 HSC/06/55. It will be seen that HSE commissioned this research primarily to ascertain whether textured decorative coatings (TC, the best known of which was Artex) presented a risk to health. Following this report all regulation of TCs was removed. This research paper also shows that no significant risk to health arises from the use of asbestos cement.

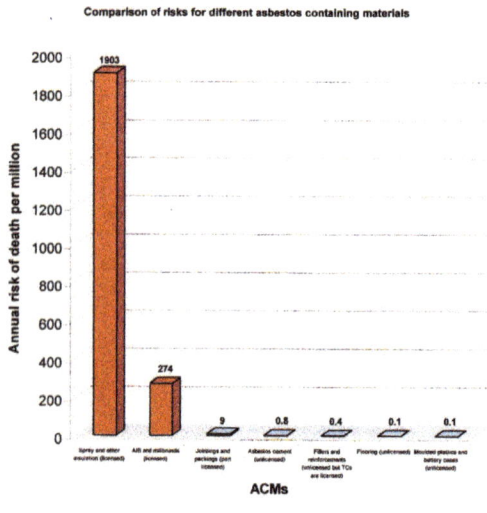

Comparison of risks for different asbestos containing materials

Figure above is based on Figure 7 from HSL's paper dealing with the risks from asbestos products (see Annex 2). It is based on an assumption of workers spending 10% of their time working with particular products for 40 years from age 20. By way of comparison, it has been estimated that licensed removal workers currently spend about 15% to 20% of their time on active removal of asbestos. No account is taken of respiratory protective equipment.

It demonstrates that the risks from work with sprays and other insulation are nearly 20,000 times the risk from TCs and the risks from AIB are nearly 2700 times the risks from TCs. The risks from TCs are comparable to the risks from other unlicensed asbestos materials.

This histogram bar chart shows the real risk of death to workers spraying insulating materials containing the amphibole forms of asbestos. However, the risk to health of workers spending 10% of their time working with asbestos cement (which contains chrysotile) from the age of 20 years up to retirement from that work at age 60 is trivial, less than 1 in one million.

Bryan with Professor John Bridle (right) and the late Christopher Booker (centre) in June 2012. Christopher Booker used to write a half-page article for the Daily Telegraph each Saturday. The subject of two of his articles, on 25th May 2008 and 15th July 2012, had been the adverse effect on farmers of the Government regulation of white asbestos (Chrysotile) cement roofing sheets. The new regulations had been instigated by the EU Waste Framework Directive 2007. The 2008 article showed a photo taken at Kensham Farm of me holding a partly broken sheet of asbestos cement, and the 2012 article was headed "Farmers to fight a £6bn asbestos scam".

In September 2012 we in the South Bucks branch of the NFU, with administrative assistance from William White, NFU SE Regional Director, hosted a fact finding tour of farms in Buckinghamshire for the Swine Raisers Association of Thailand. We showed the Thai farmers our arable enterprise at Kensham Farm, then the Lacey family's dairy and egg enterprise and finally Sally Stockings' outdoor pig unit at Ewelme, Oxfordshire. On the following day Professor John Bridle and I led the seminar (shown right of centre in this photo, both in shirt sleeves with tie) at the Lambert Arms at Aston Rowant. We showed the histogram charts from HSC/06/055 and explained how the UK Government had classified white asbestos cement as being hazardous. We then described the regulations concerning its use and disposal in the UK.

The Future Farming and Environment Evidence Compendium

February 2018

Farm Economics and Accounts

Food Production

Environmental Land Management

Department
for Environment
Food & Rural Affairs

Government
Statistical Service

In February 2018 this excellent Future Farming and Environment Evidence Compendium was published by Defra in conjunction with the Government Statistical Service.

Back to Summary | Farm Economics and Accounts | Food Production | Environmental Land Management | Supporting Information

How does profit vary across the different farm types in England?

Farm Business Income varies across the different farm types, and over the period 2014/15 to 2016/17 poultry farms were most profitable and grazing livestock and mixed farms the least.

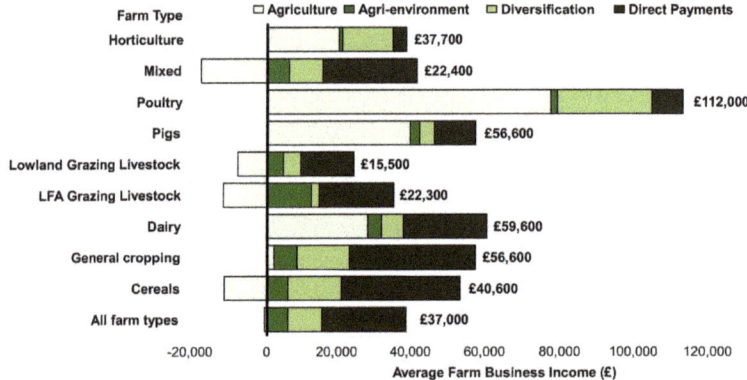

Farm Type	Average Farm Business Income (£)
Horticulture	£37,700
Mixed	£22,400
Poultry	£112,000
Pigs	£56,600
Lowland Grazing Livestock	£15,500
LFA Grazing Livestock	£22,300
Dairy	£59,600
General cropping	£56,600
Cereals	£40,600
All farm types	£37,000

Legend: □ Agriculture ■ Agri-environment □ Diversification ■ Direct Payments

Mixed, grazing livestock and cereals farms made a loss from the agriculture side of the business as their costs of production outweighed the value of their output.

Around two-thirds of Farm Business Income came from the agricultural side of the business for pig and poultry farms.

Over 75% of Farm Business Income came from Direct Payments for cereal, grazing livestock and mixed farms

The Future Farming and Environment Evidence Compendium (February 2018)   24

This chart on page 24 of the Defra Compendium of Evidence shows that without the former EU Direct payments (or equivalent schemes for farm support to replace Direct Payments after Brexit) the majority of UK farms would become insolvent and no longer able to continue farming.

An even more important Defra report was published in February 2018, entitled Health and Harmony: the future for food, farming and the environment in a Green Brexit. This set out the future direction of post-Brexit UK farming. Bryan considered the Health and Harmony policy to be fundamentally flawed by failing to treat the production of food, and with it food security for the nation, as being worthy of mention. Fortunately, the House of Lords had secured worthwhile amendments on food production before the Agriculture Act 2020 received Royal Assent.

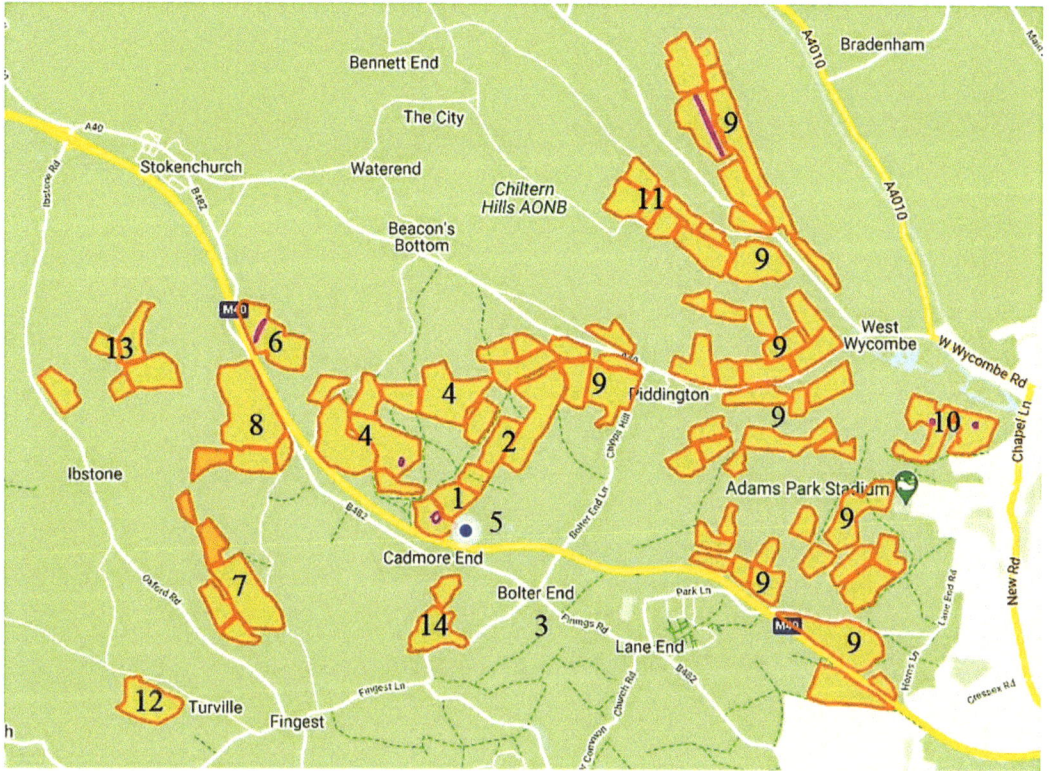

Land farmed by us family partners of the Kensham Farms partnership in 2020.

Plan prepared at Kensham Farm with John Deere Operation Centre software by driving tractor around the perimeter of the fields with a 'Greenstar' GPS receiver

Schedule of Land Farmed by Kensham Farms Partnership in 2020
Farms owned, rented on Farm Business Tenancies (FBTs) or Contract Farming Agreement

Ref	Farm	Acres	Year	Tenure
1	Kensham Farm	102	1955	Owned
2	Watercroft Farm	104	1961	Owned
3	Manor of Fingest, Common Land	25	1966	Owned
4	Dells & Bigmore Farms	240	1970	Owned
5	Manor of Cadmore End, Common Land	54	1975	Owned
6	Penley Hollies Fields	73	1985	Owned
7	Harecramp Estate	123	1990	Contract
8	Chequers Manor Top Plain Fields	112	1996	Owned
9	West Wycombe Estate at	1,331	1999	FBT
	Fillingdon, Myze, Cockshoots, Chorley, Bullocks,			
	Pyatts, Grove and Fryers Farms			
10	West Wycombe Park	73	1999	Contract
11	Chawley Manor Farm	95	1999	FBT
12	Turville Valley Farm	45	2008	FBT
13	Cholsey Grange Farm	88	2017	Contract
14	Hill Farm	60	2017	Owned

Owned by the Partnership	770	
Rented on FBT or Contract Farming	1,755	
Total Land Farmed	**2,525**	**acres**

Chapter 4

Faith in the Countryside

For the first twelve years after Alison and I started farming Kensham Farm, and then with Watercroft Farm as a 206 acre farm, my overriding concern was how to make enough profit from the farm to provide a living and a stable home for our family. During those twelve years I had little thought for anything other than ensuring the continuity of the farms by making sufficient profit during each trading year. It was not until this was achieved that I had the time to follow my wider interests, all allied to the farm but not strictly part of the farm's business.

It was in 1968 that a men's discussion group was formed by the then vicar of Holy Trinity Church at Lane End, with around eight of us men meeting regularly on a Thursday evening for discussion of matters concerning the Christian faith at the local parish church, and wider issues relating to life in general. Each evening ended with a shortened form of Holy Communion. We met weekly for nearly two years; most members of the group were in their early or mid-forties, but I was the youngest in my late thirties. All of us went on to take an active part in church life – two of our members were subsequently licensed as Lay Ministers and others were active on the Wycombe Deanery Synod discussing matters of policy and doctrine that concerned the 35 Anglican Churches within the High Wycombe area.

In my own case I joined the part-time education staff at a local Borstal penal establishment and found the work of helping those lads to sort out their own lives and future to be worthwhile and rewarding. This part-time work blended well with the continuing work of our farms.

That was the time when I was first elected to serve on the Parochial Church Council of Holy Trinity Church at Lane End, and then in 1972 to serve as one of the Churchwardens. I found it an interesting statistic, at that time, that within the Anglican Church more Churchwardens of rural parish churches were farmers than any other single occupation.

It was at the end of the 1980s that several of us from Holy Trinity at Lane End visited the parish church at Ewelme in Oxfordshire to give observations and evidence to members of the

Commission on Rural Areas in preparation for the book *Faith in the Countryside*, a report presented to the Archbishops of Canterbury and York in 1990.

Faith in the Countryside was a most comprehensive book of 400 pages, prepared by the Commission headed by the Rt. Hon. Lord Prior PC, with representatives of the Anglican, Roman Catholic, Methodist and United Reformed Churches, the Church Commissioners, four different Universities, the National Federation of Women's Institutes, the Trade Union Congress, the Council for the Protection of Rural England, the National Farmers Union, Rural Landowners and the Ministry of Agriculture, Fisheries and Food.

My own opinion is that *Faith in the Countryside*, although published thirty years ago, continues to be one of the most important books ever written on the significance of the Christian Faith, and its place in farming and rural communities. I show below four paragraphs from the conclusion of the Commissioners: -

> *"Our work has led us to see the Church as an integral part of the rural scene, not just through its history but as an active participant in contemporary life. The Church is in action in an arena in which not only local issues are focused, but also those of national and global significance.*
>
> *"We have found that beneath a seemingly tranquil surface there are great changes, full of tension occurring in most rural areas. Our studies have shown that population movements have brought new people with different perceptions and ambitions into rural life. The nature and style of work has changed, and predominantly agricultural land-use has been questioned.*
>
> *"The village hall, the voluntary transport system, the Church, and the caring network all rely on people giving of their time and expertise: the local management of schools has added another responsibility, and maybe the cottage hospital will join the list of facilities that can only be maintained if local people provide the initiative and management.*
>
> ***"The Church of England, together with its other main Christian denominations, has probably as great a stake, and certainly a more visible one, in England's rural areas as any other body."***

Finnamore Wood Borstal for Young Offenders

It was in 1968 that I first worked as a Tutor at a nearby penal establishment, Finnamore Wood Camp Borstal, for youngsters between the ages of 17 and 22.

Borstals were first developed on a national basis under the Crime Prevention Act of 1908 for erring lads who had been sentenced for 'borstal training' rather than being sent to prison. The system was designed to be educational rather than punitive but was highly regulated with a focus on regular work and discipline as well as education.

The premises of the Borstal at Finnamore Wood were a hutted camp, a few miles from Marlow, Buckinghamshire, which had been built as one of many hutted camps in the country in 1939/40 for deprived children. During World War II it was used as a school for girls evacuated from Ilford. The next use of Finnamore Wood camp was to house the rowing teams from Canada, Argentina and Brazil for the 1948 Olympic Games so that they could train on the River Thames at Marlow.

The education department, within which I was one of the tutors on a part-time basis, was quite separate from the discipline staff employed by the Prison Service. One of our tasks was to provide evening classes of one and a half hours duration for the lads to study one of around seven different subjects such as basic literacy, woodwork, decorating, electronics, and other subjects on four evenings each week. There was a full time Education Officer who interviewed each lad as an individual, to choose a suitable educational programme for his needs. The first Education Officer with whom I served was Roly Egan, who had served in World War 11 and had been taken prisoner of war – an experience which inspired him to work with young prisoners after demobilisation. I used to teach mathematics, photography, and facilitate group discussions.

It was in 1960 that Finnamore Wood became an open Borstal institution. The Criminal Justice Act 1982 had abolished all Borstals, but Finnamore Wood had continued with similar work for young offenders, initially renamed as a Youth Custody Centre then subsequently renamed as a Young Offenders Institution. As well as evening classes, all the lads worked on useful activities during the daytime. Several lads worked on maintenance of the grounds and greenhouses at Finnamore Wood supervised by Tony Falconer, the long serving, thoughtful and practical Prison Officer in charge of the extensive grounds and playing fields. Other lads worked in the grounds of a nearby Convent, and for many years we had one of the lads helping us with egg collection and feeding the hens at Kensham Farm. One afternoon each week I used to take a class of around 6 lads for tractor driving using one of our own tractors at Kensham Farm.

The education department was registered with the Royal Society for the Encouragement of Arts, Manufacture and Commerce (RSA), which at that time was an examination authority. The basic RSA exams were set at a standard that was rather lower than the former O-levels or current GCSEs used in all secondary schools. This was very suitable for some of our lads, since after only a few months of study those lads who were conscientious were able to pass an RSA exam, for which they would be awarded an RSA certificate. For many of our lads this provided them with enormous encouragement. Perhaps for some, this was the first time in their lives that they had achieved something from their own efforts, giving them that confidence in their own ability which could lead to a work ethic that would remain with them long after the months of their Borstal training.

I retired from that work after 22 years in 1990 at the age of 58 at which time my hearing had become less acute than when I was younger – a disadvantage when training lads who, in their earlier life, had taken every opportunity to act out of line with authority. In 1996 Finnamore Wood YOI was closed and sold to developers.

History of the Kensham Farmhouse

Alison and I were fortunate to have lived in just one house, the farmhouse at Kensham Farm, for the whole of our married lives from 1955 to 2020, a period of just under 65 years – although for the first year and a half of that time we had to live in one of our newly built farm cottages while the farmhouse was being renovated. This extensive renovation was necessary, since with a slum clearance order on the house we were not permitted to live in it, but as a listed building of historical interest we were obliged to preserve it.

In the *Buildings of England* by Nikolaus Pevsner and Elizabeth Williamson first published in 1960, the 1994 second edition for Buckinghamshire shows Kensham Farm as one of only three buildings listed under Cadmore End, the other two being St. Mary-le-Moor Church built in 1851 and the Brick and Tile Kiln on Cadmore End Common. The entry for Kensham Farm states: -

> *"Kensham Farm, just North of the M40. Mostly of brick and flint but with timber-framing and old brick exposed on the North wall. Inside, the 1. Bay was the 15th century open hall; what survives is a central base-cruck with chamfered arched braces to the collar, and a stop-chamfered aisle plate with curved brace alongside. The chimneystack, with diagonally-set chimneys, and the floor were inserted in the 16th century."*

In 1990 we were pleased to invite three research historians from the Tree-ring Dating Laboratory of Nottingham University to take borings from the roof timber of the farmhouse. Unfortunately, the cores taken had too few rings to analyse in the normal way, but nevertheless the research proved significantly successful for many references to be included in *"The Medieval Peasant House in Middle England"* by Nat Alcock and Dan Miles published in 2013 by Oxbow Books, Oxford, UK. From this book and from a survey and report by Barbara Wallis in 2000 and other historic and ecclesiastical records it can be seen that: -

- The earliest Kensham Farmhouse had been a hall with a central base-cruck truss in the 14th or 15th century. This construction made it unnecessary to have any aisle posts in the middle of the hall. There would have been an open fire on a hearth, with smoke rising to a gap in the roof instead of any chimney.

- John Kensham is shown in ecclesiastical records to have been the principal contributor for the Lay Subsidy for Cadmore End in 1523, and clearly his surname continues to be the name of the farm five hundred years later.

- It is thought that the chimneys were built in the 16th century enabling a substantial floor to be built, the massive timber beams which form the ceiling of the present dining room. At that time a ladder would have been used to access the new first floor with its bedrooms.

- Re-roofing had taken place in the 17th and 18th centuries. In the 20th century the existing roof members have been felted and retiled with handmade clay tiles, and in 1964 we added a new extension as a farm office with an additional bedroom and bathroom above it.

- Some of the freehold land in Cadmore End and Stokenchurch including parts of Kensham Farm's fields used to be within the Parish of Lewknor, Oxfordshire. This was known as Lewknor Uphill. Although it was transferred to Buckinghamshire in

1844 for civil purposes, it became part of the ecclesiastical parish of Cadmore End when this was created in 1852. This was the historic basis of some of our fields carrying Chancel Repair liability, a liability which I contested fiercely, as described below.

Chancel Repair Liability

Our own case of a claim against us for Chancel Repair Liability started in 1971 when Brian Ball-Green, who had recently bought Dells and Bigmore farms which he then let to us to farm as his tenant, received a letter from the solicitors acting for the Parochial Church Council of St. Margaret's Church, Lewknor, which lies in Oxfordshire seven miles away from Cadmore End, asking him to pay for roof repairs of the chancel of St. Margaret's Church on the grounds that some of the fields that he had bought recently carried the ancient liability of keeping the chancel wind and weathertight.

Brian Ball-Green, who was not best pleased with this claim, therefore asked his solicitors why they had not alerted him to any such liability when he bought the land. His solicitors then sought Counsel's Opinion, a copy of which dated 30.11.1972 is in my archives, which revealed the astonishing fact that there was no obligation on a conveyancing solicitor to make enquiries about chancel repair liability. Some of our fields at Watercroft Farm carried a proportion of the same liability, so I offered to pursue the claim against us through the church authority.

Alison and I have always been practising Christians, actively supporting the work of our own Parish Churches. We regarded this claim as being of no moral standing, even though it seemed likely that it could be enforced under ancient laws which had never been correctly resolved by the Tithes Redemption Act 1936. I therefore followed up the claim vigorously, largely to protect the reputation of the Church. The money involved was a secondary consideration, since my time and effort involved in research and lobbying was hardly justified by the amount of money claimed from us.

I am pleased that our efforts were eventually successful, albeit that it has taken thirty years for the laws of England, regarding registration of chancel repair liability with the Land Registry, to be changed. It is now standard practice for all conveyancing solicitors to make formal enquiries about the liability from the vendor's solicitor.

In 2010, I prepared a guide on the subject for the website of our own parish church, Holy Trinity at Lane End. I started the guide by pointing out that at that time I had been Churchwarden of Holy Trinity Lane End and at one time I had been a reluctant "Lay Rector" of St. Margaret, Lewknor. I acknowledged with gratitude the very considerable encouragement and assistance which I had received over the years from the Rt. Revd. John Bone, sometime Archdeacon of Buckingham and later Bishop of Reading.

My research showed that within National Records: -

- *No Schedule of Chancel Repair Liabilities existed for the Churches of England and Wales nationally.*

- *HM Inland Revenue attempted to prepare a schedule many years ago. This Schedule was named the "Record of Ascertainments 1940", but within it there were errors and omissions. This has been retained by The National Archives on microfilm.*

- *The National Archives Legal Information Leaflet 33 is an excellent summary of the chancel repair liabilities which benefit some 5,200 pre-Reformation Churches in England and Wales.*

In February 1982, at my instigation the General Synod of the Church of England held a debate on the subject. The then Archdeacon of Buckingham (later Bishop John Bone) spoke in favour of calling for the abolition of Chancel Repair Liability which he described as being 'arbitrary, inequitable and archaic'. Synod passed a Resolution, recommending that Chancel Repair Liability be phased out over a period of years, and that this decision be passed to the Law Commission.

The Law Commission then prepared its Working Paper No. 86 (Green Paper) entitled *'Transfer of Land Liability for Chancel Repairs'* which was published in June 1983. The author was one of many respondents, all of whom are listed in Appendix C of the subsequent Law Commission (White Paper) Law Com. No. 152 published in November 1985 and entitled *'Property Law, Liability for Chancel Repairs'.*

The White Paper included a draft Bill intended to "Make provision for ending the liability of Lay Rectors for the repair of chancels" after the end of 1995, thus recommending a 10-year period in which Parishes could ensure that their Lay Rectors had carried out their responsibilities of maintaining the Chancels for which they were responsible in "wind and weathertight condition".

However, this legislation was never enacted, perhaps the result of insufficient political pressure or perhaps because it could have been construed as inequitable that Parishes might have lost, without compensation, the benefit of a Lay Rector's funds to which they had been previously entitled for the repair of the Chancel.

No action was taken until 1994 when the Wallbank Family lost a widely publicised case brought against them by the Aston Cantlow PCC for repairs to the roof of the Chancel of St. John the Baptist Church at Aston Cantlow, Warwickshire. After two decades of legal wrangling when the case had cost the Wallbank family a total of £600,000 in legal fees, the family were forced to sell their Warwickshire farm.

Government then recognised that Chancel Repair Liability was a serious flaw in conveyancing procedures of properties within England and Wales which could affect the owners of private houses and their gardens as well as farms and estates.

As a solution to this flaw, Government inserted an additional clause within the provisions of the Land Registration Act 2002 which required all Chancel Repair Liabilities to be registered at HM Land Registry before the end of September 2013. The effect of this is that now, in 2021, both Parochial Church Councils and property owners and farmers know that if there is mention of a Chancel Repair Liability under the Title Number of the property at HM Land Registry then such a liability does exist. However, if there is nothing on a Land Certificate mentioning Chancel Repair Liability then the owner will know that no such claim can be brought against him.

Construction of the M40 Motorway over Kensham Farm

When my father bought Kensham Farm (102 acres) in 1946 he was aware of the route of the High Wycombe Bypass which subsequently became the route of the M40 Motorway.

My recollection is that then we did not hear anything further until the early 1960s - when we received by registered post a large envelope with draft plans, using the original Bypass route as the proposed site for the M40 and a letter saying that if we did not voluntarily sell the land required for the new M40, then it would be taken from us by compulsory purchase. At that stage we made extensive enquiries into the past history of this proposed road, and the subject of motorways in general.

A meeting was set to take place in St Mary le Moor Church Village Hall where those of us objecting would be able to have our opinions heard. Prior to that meeting Alison and I invited Barbara Castle, at that time a Labour MP before she became Minister of Transport, and others from the village to a private discussion in our dining room. We were shown plans for four alternative routes, but I never had copies of them. We worked out a plan for the local public meeting in Cadmore End, which was to press for the route to go through Princes Risborough.

When the public meeting was held, we were given logical reasons why the Princes Risborough route, and the two other alternatives, would not be viable. Barbara Castle (who attended in her capacity as a private citizen, owning 'The Beacons' in Cadmore End as a second home) then said, "Well, if it has to come near Cadmore End, why cannot it go through the middle of Kensham Farm instead of across one corner?" My recollection is that I gave a robust explanation of the reasons why I thought that would not be a good idea.

Then, after the route was chosen, largely as planned in the 1930s, Barbara Castle sold her holiday house in Cadmore End and moved to Ibstone. We received compensation of £1,600 as negotiated by Savills acting for us - the computation was £200 per acre for the freehold of six acres taken, with £400 for the 'Injurious Affection' which included leaving approximately six acres on the wrong side of the proposed new M40 (i.e., the village side), and for all other counts for all time. This money was inadequate to cover the purchase of any replacement land elsewhere, even at that time.

The other interesting point concerned Cadmore End Common. Even the Minister of Transport is not allowed to compulsory purchase Common Land without replacing it with other ordinary land adjoining the Common. The Minister then had to decide who owned the Common (such matters were clouded in the mists of antiquity before the Commons Registration Act 1965) since the late Ronald Williamson and the Edgley Family were in dispute over ownership of Cadmore End Common. After further historic research the Minister decided that the Edgleys owned Bolter End Common and the eastern parts of Cadmore End Common, whereas Ronald Williamson was deemed to own the western part of Cadmore End Common over which the M40 would be built. Subsequently we (Edgleys) bought the remainder of Cadmore End Common from Ronald Williamson's widow, Dorrie, when she moved to Dunstable.

The necessary additional farmland chosen by the Minister to become part of Cadmore End Common had been part of a small field known as Hatches Meadow (which the late Norman Lacey of Rackleys Farm used to graze with cattle), just south-east of Hatches Pond on Cadmore End Common.

At that time, Hatches Meadow had been within the land which had included New Road, Bolter End, a commercial sand quarry behind the Peacock Pub (when we first farmed Kensham Farm I could buy building sand there and collect it in the Land Rover) and a good pond known as Marlins (used for recreation by many local lads) near the present-day Lane End Parish Council Allotments. A significant stretch of this land was used for building the new motorway, in addition to the part that has become part of Cadmore End Common since construction of the M40.

The actual building of the M40 was completed in stages. When it was originally designed as the High Wycombe Bypass, it was intended to improve the A40 as part of the London to Fishguard trunk road. However, when the M4 was completed first, including the Chepstow Severn Bridge, this became the main route to Wales. Thereafter the main purpose of the M40 became to connect London and Birmingham.

The first stretch of the M40 to be built was from junctions 4 to 5, High Wycombe to Stokenchurch, which cut off the south-west corner of our Kensham Farm land including Kensham Cottages, was opened in 1967. Junctions 1 to 4 were built after that and the northbound section from Stokenchurch to Oxford was completed in 1974. Subsequently in 1989 the width of the motorway was increased from two lanes to three, prior to the M40 being extended to Birmingham. This work involved the building of a temporary bridge to provide access to Kensham Farm during the months when the original bridge was being demolished, so that it could be replaced with a wider span bridge.

The National Farmers Union

When I joined the NFU in Autumn 1955, the average size of farm was far less than nowadays, so more NFU branches were necessary for the larger number of farmers, most of whom were members of the NFU. However, the need for farmers to influence political decisions was as great then as now.

The NFU branch which I joined was the High Wycombe Branch, representatives of which sat on the Buckinghamshire County Executive. I was first elected as High Wycombe Branch Chairman in 1961, when I was 29, and I attended the monthly meetings of the County Executive to represent the branch. When I joined Sir James Turner, later Lord Netherthorpe, was the NFU National President, but by 1960 he had retired to be succeeded by Sir Harold Woolley. There have been nine Presidents since then; the current President, Minette Batters, who is the first woman to have ever been elected, has been outstanding at recognising the problems facing the farming industry which are likely to affect the viability of the average family farm in the years to come.

The structure of the NFU governance has evolved sensibly in line with the economic forces which have caused so many farms to amalgamate into larger farm units. This has been necessary to remain viable as producers of the raw ingredients of the food which most people buy in supermarkets. The NFU moved its headquarters from Agriculture House, Knightsbridge, which had been opened by the Queen Mother in October 1956, to its present headquarters at Stoneleigh Park, Warwickshire in December 2005.

When my Alison first bought our household provisions it was from John Carr's grocery shop in Lane End village, where John or one of his assistants would take the cans or packets of fresh food off the shelves and place them on the counter ready for the customer to pay or put onto a monthly account. And if bacon was required by this customer, then John would personally cut the rashers off the side of bacon with his electric slicing machine. That was a time when food was often locally produced and bought by the retailer for sale to the customer.

Supermarkets then took over from the smaller retailers and required supplies of raw food ingredients on a national scale. Thus, food production from the farms has had to evolve and be offered wholesale in quantities appropriate for a national food marketing corporation. It has been interesting in more recent years that Farm Shops have become more popular – reverting to the scale of local produce to be sold locally by a personal, generally family-run, farm shop.

My second stint of the two years as Chairman of our local branch was in 2011, precisely fifty years after the first stint, but by then the local branch was no longer High Wycombe, it was the whole of South Buckinghamshire and Middlesex, which itself forms part of the County Branch of Buckinghamshire, Berkshire and Oxfordshire. This together with Hampshire, Kent and West Sussex forms the NFU South East Region with an office in Petersfield. It has been my delight at the Branch AGM in Autumn 2019 to see my grandson, Alex Nelms, elected to the same position as Chairman of the South Bucks and Middlesex Branch.

I served on the Southeast Regional Poultry board of the NFU for many years until we closed the egg production side of our farming activities in 1992. However, my most significant work for the NFU has been since March 2001. One of the first briefing papers that I prepared was at the request of Mrs. June O'Dell OBE, the Chairman of Business and Professional Women UK Limited, for the annual conference of BPW at Manchester in April 2001.

Some points made in my BPW Briefing Paper

I drew attention to the low ebb that farmgate production prices had reached. I was able to point out that inflation between 1971 and 1999, a period of 28 years had been quite astonishing, such that average prices had increased by 756%. However, after making allowances for that inflation, most everyday requirements had increased at a greater rate than inflation, but the cost of food had diminished: -

The farmer received only 7 pence for the wheat from which a loaf of bread is made retailing at an average price of 63 pence.

Item	1971 cost	Index linked inflation to 1999	1999 actual cost	% difference actual vs index
3-bedroom house in SE	£8,000	£60,532	£99,903	+ 65%
Single State Pension	£6.00	£45.40	£64.70	+ 43%
Gallon of petrol	33.5p	£2.53	£3.04	+ 20%
Loaf of bread	9.5p	72p	63p	-13%
Pint of milk	5.5p	42p	27p	- 36%

At that time the England Rural Development Programme (ERDP) set out by the Ministry of Agriculture, Fisheries and Food in the Government's November 2007 Rural White Paper had failed to put food production first. The Women's Institute had expressed its views in 'The World of W.I.' HQ News that:-

"At a time of growing climate, ecological and global economic instability, it is a high-risk policy that threatens to let our best agricultural land be turned over to development, our best farmers be turned into tourist guides and our nation becoming increasingly dependent on imported foods brokered by giant transnational corporations.

Rural policy is, of course, about much more than agriculture, but agriculture is the linchpin. How we obtain our food and manage our land and natural resources is central to our economic, social, cultural and ecological health."

Berkshire, Buckinghamshire and Oxfordshire (BB&O) County Branch of NFU Strategy Guidelines for UK Agriculture following the Curry Report

My next significant work for the NFU was assisting David Orpwood, at that time Chairman of the BB&O NFU branch, to prepare a new 'Strategy for Agriculture' for submission to NFU HQ following the Curry Report.

Sir Donald Curry had introduced his Report entitled *"Policy Commission Investigation into the Future of Farming and Food"* to the AGM of the NFU which was held in London on 7th February 2002. In his speech Sir Donald pointed out that the farming crisis was such that tinkering at the edges would be insufficient to bring sustainability back into British farming. He pointed out that profit is the key, but that an increasing proportion of the cost of food paid by the consumer goes to other players. Sir Donald wanted to see improved marketing, including country-of-origin labels, and advised that Government should make a clear statement of its support for farming as a sustainer of the rural environment as well as food producer. He said that profit and environment are linked, and that the NFU should take a lead in bringing back pride and trust in the farming industry and reconnecting it with consumers and the countryside.

Bearing in mind Sir Donald's comments in his Policy Commission Investigation, David and I produced a full report urging NFU Headquarters to publish a new Strategy Document. This was approved by the BB&O County Branch in the following Resolution: -

> ***"This Berkshire, Buckinghamshire and Oxfordshire County Branch of the National Farmers Union firmly believes that there is a future for British Agriculture. To this end we call on NFU HQ to publish, without delay, a 'Strategy for UK Agriculture – 2002' to recognise the achievements of British Agriculture and to resist any further export of our food production capacity to overseas countries".***

Most of my subsequent work for the NFU concerned Environment Agency matters, in particular the regulations concerning the disposal of Chrysotile asbestos cement roofing sheets. However, we hosted the important Farmer Engagement Meeting in February 2018 in our dining room at Kensham Farm with three Defra officers and fifteen of us farmers to discuss the impractical proposals for the future of British agriculture set out in the Defra *Health and Harmony* consultation paper.

Three months later, in May 2018, we hosted a similar meeting at the request of the NFU to explain to the French Ambassador to Britain, Monsieur Jean-Pierre Jouyet, what farmers in Britain thought about the Defra proposals for future British agricultural policy when independent of the EU. We explained our views to the ambassador openly, that agricultural policy should be centred on the production of food, whereas the *Health and Harmony* proposals attached little importance to food production. We felt that the Ambassador was likely to return to London with a favourable impression of rural folk in Buckinghamshire, since the NFU had arranged for him to go on to the Rebellion Brewery at Bencombe Farm, Marlow Bottom to complete his tour!

Regional Environmental Protection Advisory Committee (REPAC) for the Thames Region

In 2003, I was appointed as the farming industry representative on the Environment Agency Regional Environmental Protection Advisory Committee (REPAC) for the Thames Region.

Prior to my appointment, the Chairman of this committee, Pamela Castle FRSA, was approached by Ben Gill, at that time President of the NFU, at a reception where she told him that there ought to be a farming representative on the REPAC Committee. Ben Gill then asked the Regional Director of the NFU, Shaun Leavy, to find a farmer willing to serve, so he approached me – it was more of an order rather than a request to serve on this Thames Region REPAC, but I was pleased to agree. I attended all the quarterly meetings of the Thames Region REPAC at Reading until most quangos, including all REPAC Committees, were disbanded by the then Prime Minister, David Cameron, in 2012.

When I was first appointed to the Thames Region REPAC, I found that the Chairman of the Environment Agency Board in his EA Annual Report for 2002/3 had claimed that British Agriculture caused more harm than good to the environment. I was able to take up that false contention, with both the Chairman and Chief Executive, by suggesting to them both prior to one of the EA Board meetings at Reading Town Hall in 2004 that if they were correct, then the environment would improve if the whole of British Agriculture were to be closed down. We then had a useful discussion on the subject, following which the Agency treated the broad benefits of agriculture in Britain with respect, without detracting from its work of monitoring and encouraging improvements over such matters as the safe disposal of sheep dip and other chemicals which could be harmful if allowed to contaminate watercourses.

Another aspect of the Environment Agency's work over which I pressed for improvement was its earlier policy of engaging with farmers both as adviser and as policeman and prosecutor. I urged the Agency that officers at District level should be regarded by farmers as useful advisers, and that it should only be farmers who had ignored earlier advice from EA District Advisory Officers who should be prosecuted for any environmental offences. I made it clear that any prosecution should be by completely different staff at National or Regional level only.

Concerning the broad attitude of the EA towards the British farming industry, I was pleased to see in a subsequent Executive Summary of the report by the EA Director of Environmental Protection in 2004 entitled *'Agriculture and Natural Resources: Benefits, costs and potential solutions'* that benefits of agriculture included: -

	Environmental Benefits		Social Benefits
1.	Aesthetic value of Agricultural Landscapes	9.	Provision of jobs
2.	Recreation and amenity	10.	Contribution to the rural economy
3.	Water accumulation and supply	11.	Contribution to rural communities
4.	Nutrient recycling and fixation		
5.	Soil formation		
6.	Wildlife protection		
7.	Storm protection and flood control		
8.	Carbon retention by trees and soil		

Chrysotile Asbestos Cement

It was in 2007 at a meeting of the Thames Region REPAC that we discussed the new EU Waste Framework Directive and in particular the section concerning proposed new regulation that would seriously affect the very large number of farms that had post World War II farm buildings with asbestos cement roofs. Before the EU Waste Framework Directive, asbestos cement had always been regarded as the most suitable roofing material for agricultural buildings and barns. It was stocked and used by all the manufacturers and erectors of farm barns and by many builders' merchants. There had been no controls or even any thought that one day the EU might classify it as a hazardous material potentially harmful to those working with it or using it.

I realised the huge significance this would have for the agricultural industry and so I arranged a conference at NFU Agriculture House, Eynsham in April 2008 for the purpose of investigating the science of the subject. I wanted to have three principal speakers and invited Professor John Bridle who had been involved in the asbestos industry for many years, and Doctor John Hoskins, sometime Chairman of the Toxicology group of the Royal Society of Chemists, who would explain the science of the subject. I invited the Health and Safety Executive (HSE) to appoint a third speaker to explain the regulatory side of implementing the new EU requirements set out in the Waste Framework Directive. It was a great disappointment to me that my request for a speaker was rejected on the grounds that the HSE was not willing to share a platform with these two speakers.

Doctor John Hoskins explained that the term asbestos was a loose term which included Chrysotile, a magnesium silicate normally called 'white asbestos' which, when combined with cement to make asbestos cement roofing sheets, had never been shown to have caused any ill health. However, the term asbestos also included Amosite normally known as 'brown asbestos' and Crocidolite normally known as 'blue asbestos' which at an earlier stage had been mined extensively in Australia. These amphibole forms of asbestos, Amosite and Crocidolite, were iron silicates containing small needle like fibres which are insoluble in acid. The risk to human health from the amphibole forms of asbestos had already been recognised and controlled by legislation, with most mining of the amphibole forms having been already closed before the EU Waste Framework Directive was introduced.

The dangerous amphibole forms of asbestos had been used extensively before and during World War II as pipe lagging and insulation for boilers where its properties of being a good insulator that would not be harmed by any acid were most useful. At that stage it had not been realised that amphibole fibres in the air could be breathed in, especially by workers such as plumbers and operators of spray insulation plant, and not suffer any ill effects for many years. However, the fibres could remain in the lung and eventually, perhaps forty or fifty years later, penetrate the lung and irritate the pleural lining that surrounds it thus causing the most painful and fatal condition of mesothelioma.

Chrysotile, the white asbestos used in asbestos cement, consists of soft fibres of magnesium silicate which are a useful binding agent for the cement for roofing panels performing the same job as mild steel rod in reinforced concrete. Because these fibres are soft, and soluble in acid, even if they were to be unfortunately breathed in by a worker, the fibres will be dissolved out within a relatively short period by the natural acid in the lung.

Our campaign for recognition that Chrysotile asbestos cement did not share the harmful properties of the amphibole forms of asbestos was then followed up by Christopher Booker who at that time was a regular contributor to *The Sunday Telegraph*. He published an article on 15th July 2012 in which he pointed out that powerful lobby groups representing health claim lawyers and asbestos removal contractors had managed to hijack government policy by attributing the known harmful effects of the amphibole forms of asbestos to Chrysotile asbestos cement, which had never been known to cause ill health. He drew attention to my application for a Judicial Review to challenge the provisions of the 2011 regulations concerning the use and disposal of Chrysotile asbestos cement. Regrettably my application was turned down in August 2012.

In all my work over a period of fourteen years on the regulation of asbestos cement I had relied heavily on the findings of the research carried out by the Health and Safety Commission. The Commission's paper HSC/06/55 published in July 2006 entitled *'A comparison of the risks from different materials containing asbestos'* showed, in a series of histogram bar charts, the very considerable risk of death to workers who had been using spray forms of insulation compared to the relatively negligible risk to workers who had been using Chrysotile asbestos cement.

The current Control of Asbestos Regulations (CAR12) followed the guidance specified in an EU Directive. Perhaps, following Brexit and the end of EU regulation, some changes may be made to CAR12 to make them more realistic and user-friendly to farmers whose barns are roofed with asbestos cement sheets.

Wycombe District Council, Farm Diversification Policy

In 1986, at a time when farm profit margins had been falling, I was approached by a local builder, John Cook, and an upholstery student at the High Wycombe College of Technology and Art, Tony Kingston, each of whom required space for workshops to carry out their respective trades. We provided them with suitable small units in the farm buildings at Watercroft Farm which were not currently being used for agricultural purposes. We prepared formal Licences to Occupy in which I made it clear that there were no sanitary facilities at Watercroft Farm and they each had to provide their own Elsan chemical toilet. I fully realised that neither of them did that, but chose instead to go behind suitable bushes. However, these uses of farm buildings for trades other than farming were successful with no detriment to our farming. I therefore asked our Agent to approach Wycombe District Council, without mentioning the name of the client who had made the enquiry, as to whether we could be granted planning consent to reschedule the use of some of our redundant farm buildings from Agriculture to Craft and Light Industrial Units. The Agent was told by the Wycombe District Council Planning Authority that such a request would certainly be turned down and that therefore they would not waste any officer's time in making a site visit to the farm buildings in question.

With my Alison and my son Charlie, who had only left the Royal Agricultural College at Cirencester a few years earlier, we decided that there was a strong moral as well as practical case, in addition to considerable local demand for small workshops, for using redundant farm buildings in this way. We therefore decided to go ahead in converting the Watercroft farmstead to Craft and Light Industrial Units, with new sanitary facilities for handwashing and WCs. I then sent a report to our local elected District Councillor Paul Ensor, who I knew well, enclosing copies of our proposals and our Agent's Report of his discussion with the Wycombe District Council Planning Authority. I asked Paul to file my letter and explained to him that we were starting without Planning Permission and were fully prepared to meet the Council in Court if that became necessary.

We then renovated the farmstead at a cost of some £70k with new roofs and appropriate partitions and repairs where necessary. Prior to advertising locally that these workshops were available, we had no difficulty in finding suitable Licensees for all ten of the Units. These included John Cook the builder, another two joinery firms, a precision engineer, a picture framer and restorer and the upholsterer Tony Kingston. All of these small businesses were well aware that no planning permission had been granted for a change of use and this was all recorded on their agreements for Licences to Occupy.

Our traders had been carrying out the work of their businesses satisfactorily without causing any local aggravation or difficulties for two years until one day we received enforcement proceedings from the Wycombe District Council to close all the units down and return them to their former agricultural use.

We lodged a formal objection to the enforcement order and engaged the services of our own solicitor at Henley-on-Thames and a Planning Officer and a Barrister suggested by our solicitor. At the subsequent three-day Public Inquiry in the Council Chamber of Wycombe District Council, the Planning Inspector who was a qualified architect in addition to being an inspector, heard all the arguments including the response from the Council official as to what the Council hoped would happen to the redundant farm buildings in question. The Inspector was not best pleased with the Council Officer's answer that he hoped the buildings would deteriorate and eventually fall down. It was at that point that we realised we had convinced the Planning Inspector that our conversions to Craft and Light Industrial Workshops were a suitable use for redundant farm buildings, and so our objection was upheld at the Inquiry and all our Craft Workshops continued their respective trades without further hinderance.

At no stage was there hostility between Wycombe District Council and me, it was merely a case that the Council had been trying to uphold National Policy over the use of farm buildings. I was particularly pleased in the years following the Inquiry that the Wycombe District Council on two occasions asked me to give illustrated talks on the subject, since by then the re-use of redundant farm buildings for trades other than farming had become standard and approved practice.

It was at the end of the 1990s that Councillor Jean Gabitas of Wycombe District Council felt that there were no longer any farmers who were District Councillors, and that therefore there should be a Rural Forum held by the Council so that the voice of farmers and others in rural areas could be heard. I was one of the founder members of the Wycombe District Council Rural Forum which was then formed, with a structure of meetings in each March and October in the Council Chamber with the meeting chaired by the Chairman of Council. All District Councillors and representatives of the farming community and parish councillors have always been invited to attend, with each year a Farm Tour being laid on by us farmers in the late afternoon in June or July to a different farm each year, with refreshments and time for informal discussion following.

Royal South Buckinghamshire Agricultural Association

The Royal South Bucks Agricultural Association (RSBAA) was founded in 1833 for the encouragement of industrious labourers and servants. It is one of many local agricultural associations which runs a ploughing match and show. The RSBAA Ploughing Match is held annually on the first Wednesday of October followed by a lunch in a marquee in the ploughing field for the members and ploughmen. Normally around four hundred members and guests sit down to a three-course lunch served from a field kitchen.

In the history of the RSBAA, there is a delightful report that: -

> *"In the latter part of the year 1833, our esteemed Vice President G.S Harcourt Esq. and other gentlemen interested in Agriculture, feeling the great want of something to stimulate Farm Labourers and Servants to greater industry and skill in their several callings, and to increase respect for moral character, met at Salt Hill. They resolved to establish a Ploughing Match, and to distribute awards among the deserving, with a view to realising the great desideratum. Exertion was made and 73 of the neighbouring Noblemen, Gentlemen and Farmers came forward; the sum of £64. 14s 6d was subscribed; a Ploughing Match was held on the 3rd December in a field kindly lent by William Nash Esq., of Langley, where 19 prizes and rewards were given away; and thus 'The South Bucks Agricultural Association' was formed."*

I was so pleased that for the two years 2017-18 when I was elected as President, my Alison was by my side. The RSBAA always uses judges from a neighbouring county's association, and it was a pleasure for me at the lunch after the 2018 Ploughing Match to hear that my son

Charlie had won the prize for the Best Large Farm and that it was my job to present the silver trophy cup to him.

In 2017 our Annual Ploughing Match at Lord Parmoor's Estate at Frieth was recorded for the television programme *"The Farmers' Country Showdown"*. The programme concentrated on the horses ploughing at the match, and on the preparation of the horses at the farms or smallholdings where they were kept. However, I and the RSBAA Secretary Andy Hall and others, were interviewed at the ploughing match. It was a delight when a few months later one of my former Borstal lads, from whom I had not heard for nearly forty years, phoned me to comment on the programme and to discuss old times when he was an internee at Finnamore Wood camp, and on his success in the hotel trade in the ensuing years.

In September 2020 the last job carried out by our Foreman, Nigel Rogers, was to harrow, roll and assist with the seeding of the new airfield on our Chequers Manor Top Plain land for use by the High Wycombe and District Model Aircraft Club. The RSBAA Secretary, Jo Short, has confirmed that Nigel's 58 years and 7 months of employment with me at Kensham Farm has equalled the previous longest ever Long Service Award which had been granted to James Lesley, a farm labourer living in Jordans Meeting House, Chalfont St Giles, in 1882 in recognition of 58 years of service.

Chapter 5

Reproduction of the Author's column in The Clarion entitled 'On the Land' from Spring 2007 to Spring 2021.

The Clarion *is the Parish magazine published by the Lane End Parish Council, with 1,750 copies being distributed, free of charge each quarter, to all the homes within the villages of Bolter End, Cadmore End, Lane End, Moor End and Wheeler End near High Wycombe, Buckinghamshire*

Winter	*2010*	– *Excess of Rainfall in August. Ventilated Floor Grain Drying. Continuous Flow Grain Dryers. Russian Grain Harvest. Wheat Exports to Morocco*
Spring	*2011*	– *Sustainable Farming Practices in the Chiltern Hills*
Summer	*2011*	– *Forestry Commission Woodland. Chilterns Commons. Purchase of Ditchfield Common by Lane End Parish Council*
Autumn	*2011*	– *Heatwave in Early Summer Months. Market Value of Grain. Grain Sales by Openfield, a Farmer Owned Cooperative. Chinese Lanterns*
Winter	*2011*	– *Michaelmas Day. Solar Photovoltaic Panels. Blackgrass and Sterile Brome Grass*
Spring	*2012*	– *Farm Work in Winter. World Grain Trade. Imports of GM Soyabean. Genetic Engineering. Chilterns Woodland Project*
Summer	*2012*	– *Irrigation and Abstraction Licences. Saddle Gall Midge Larvae. Oilseed Rape*
Autumn	*2012*	– *Milk Supply Chain. Farm Subsidies. Set-Aside Land Brought back into Cropping*
Winter	*2012*	– *Spread of Bovine Tuberculosis. Compulsory Slaughter of Cattle infected with Bovine Tuberculosis. Trial Cull of Infected Badgers*
Spring	*2013*	– *Rainfall Records. Difficulties in a Wet Autumn. Felling Roadside Trees. Mobile Sawmill*
Summer	*2013*	– *Beamish Living Museum of the North. Seed Drill Mechanism. Poor Establishment of Wheat in a Wet Autumn*
Autumn	*2013*	– *Combine Harvester. Arable Crops*
Winter	*2013*	– *CAP Reform. MacSharry Reforms. The Single Farm Payment. Pillar 1 and Pillar 2 Payment*
Spring	*2014*	– *RASE Medal for Nigel Rogers. Plant Protection Chemicals. Specialised Farm Production. Flooding on the Somerset Levels*
Summer	*2014*	– *Pastures Damaged by Flooding. Benefit of a Dry Harvest and Autumn. Open Farm Sunday. New Grain Dryer and Grain Store*
Autumn	*2014*	– *Michaelmas Day. Early Harvest. Variable World Price of Wheat. Incorporation of Straw into Soil*
Winter	*2014*	– *Berkshire College of Agriculture*
Spring	*2015*	– *Spring Drilling. Vernalisation of Winter Wheat. Importance of Good Seedbed. Malting Barley*
Summer	*2015*	– *Growth Regulator for Cereal Crops. Importance of Politics for the Farming Industry. DNA Testing of Badgers. Food Production or Conservation?*

Autumn 2015 – Wheat Harvest. Computer Monitoring on the Combine Harvester. International Trading of Wheat as a Commodity. Diamond Wedding

Winter 2015 – Crop Rotation. Development of Hormone Herbicides. Use of Glyphosate

Spring 2016 – Winter Work on the Farm. Future of UK farming if we leave the EU. Farm Gate Prices for Farm Produce. Care of the British Countryside. Badgers and Bovine Tuberculosis

Summer 2016 – Open Farm Sunday. Roofs of Farm Buildings. Risk Analysis HSC/06/55. Different Types of Asbestos

Autumn 2016 – Brexit Referendum. NFU Key Aspects for Regulation of British Farms. Harvest 2016

Winter 2016 – Currency Rates of Exchange. Mustard Plants as Game Cover. Importance of a Good Seedbed

Spring 2017 – Methods of Farm Support. Decline of British Farming in the 19th Century. German U-boats in 1914. World War II Food Rationing. Agriculture Act 1947. BCA Student Visit. Oxford Diocesan Plough Wednesday

Summer 2017 – Rainfall Records. Mains Water Supply. Johne's Disease. Portakabin Crane Hire

Autumn 2017 – Politics and Shortcomings of the CAP. Start of Harvest. Farming Calendar

Winter 2017 – Old Year Quarter Days. Stability of British Food Production. RSA Food, Farming and Countryside Commission. Royal South Bucks Agricultural Association

Spring 2018 – Soil Temperature. Fungicides and other Crop Treatments

Summer 2018 – Aerial Photography with a Drone. Start of Spring Seeding. 'Health & Harmony' Defra Consultation Paper

Autumn 2018 – Drought of Summer 2018

Winter 2018 – Ploughing Match. Post-Brexit UK Agricultural Policy. Disastrous Agriculture Bill of 2018

Spring 2019 – Food Shortages if 'No Deal Brexit'. Food Security. Spring Drilling

Summer 2019 – Balance between Food Production and Care of Countryside. Control of Couch Grass. Work of Agricultural Research Scientists

Autumn 2019 – Harvest 2019. Farming Calendar. Brexit and future of British Farming. RSA Food and Countryside Commission

Winter 2019 – Crop Rotation. UK Agriculture since World War II. Autumn Seeding. Brexit

Spring 2020 – Agriculture Bill 2020

Foreword

I first met Bryan and Alison around sixteen years ago when we moved to Lane End. It was our first grown-up house after several years of living a fast-paced city lifestyle in London flats. I was determined to embrace village life and, whilst not being religious myself, accompanied my husband to the village church around the corner from our house. I was keen to meet people outside our peer group of London escapees and new parents.

Bryan and Alison were established members of the church, in fact Bryan was in his second stint as Churchwarden at that time. They introduced themselves between the pews and my main memory of meeting them was of overwhelming kindness and welcome. Bryan twinkled paternally and Alison asked interested questions about what brought us to the village, offered us advice and generally bathed us in her unique glow.

From that day, they continued to greet us warmly whenever we saw them in the village and it made us feel accepted. It was an incredibly sad day last year when we stood with so many others to pay our last respects to that wonderful lady as the hearse passed through the streets to her beloved church. Covid had robbed us of the opportunity to say goodbye at the funeral, but the number of people huddled in their socially distanced groups along the roadside was testament to the number of lives she touched.

I understand that Bryan and Alison were devoted partners in every aspect of their life together for 65 years. Losing her must have been an unimaginable loss for Bryan, but what an amazingly positive reaction – to write a book! It's been fascinating to read about their early life together and the way that farming – and life – has changed since the 1950s.

I knew that Bryan was a good writer as he's written a column in The Clarion, the parish magazine I put together for Lane End and local villages, for as long as I've been editor. In fact, I inherited him when I took on the magazine in 2010. I was an editor of architecture and design magazines by profession, but The Clarion was (and still is) a labour of love for a nosy village-life observer like me.

Bryan's column 'On the land' stands out amongst the schools' news, WI reports and gardening advice. It's not something you can skim, you need to properly engage with it. There's information and reasoned argument to digest. One aspect of his writing that I both admire and find amusing in an affectionate way, is his precision.

Bryan would never just talk about planting wheat with a tractor. He would explain that they use a 'Weaving drill mounted on the hydraulic linkage of a 215-horsepower tractor requiring a width of 6 metres. The hopper holding 1 tonne of seed with a spot rate of around 12 acres per hour.' I have no idea what half of it means, but as a reader, I just feel honoured to be trusted with that level of detail.

Over the years, I've learned so much about what it takes to coax food from the land as a farmer – as have the rest of the residents of our villages. In these days of supermarkets and pre-packaged food, it's an invaluable insight. Whilst we only correspond every few months for The Clarion and will pass the time of day when we meet in the village, I do feel privileged to call him my friend.

Katy Dunn
Editor of *The Clarion*
August 2021

The Clarion - *Spring 2007*

The Case for locally Grown Food

Wheat growing at Watercroft and Dells Farms in mid-summer.

Following the last issue of The Clarion there was a suggestion that many folk within The Clarion territory often viewed and drove past the farmed fields in the area, but did not know a lot about the farming processes that were taking place nor the food that was being grown. So the objective of this snippet, and hopefully of future short commentaries in The Clarion, is to put the case for the food that is grown locally and for the important by-product of the local farms which is a landscape and countryside kept tidy and in good heart. A house that is empty soon becomes derelict, and similarly fields that are not farmed could soon grow scrub and look as though they had been abandoned.

Winter Work on Arable Farms

What have farmers been doing in the winter? On the arable farms most of the fields that grow crops were cultivated and seeded in September or October and any further tractor work would have done more harm than good when the land is as wet as it has been during the past winter. So for these farms, on which wheat for milling into flour for bread and biscuits is one of the main crops, the winter has been a time for checking the grain stores, loading lorries with wheat for the mills, trimming hedges, maintaining the buildings and access tracks, controlling vermin, and overhauling the farm machinery which will soon be needed for field work in the Spring.

Winter Work on Dairy Farms

But for livestock farms the Winter is a busy time, with livestock in yards where every day food and bedding straw must be carted in to the stock and the manure must be carted out. On dairy farms the cows must be milked twice each day and those farmers have had a hard time financially following the demise of the Milk Marketing Board. The wholesale price of milk at the farmgate has fallen, but the cost of milk to the housewife has increased as a result of the supermarkets taking more than their fair share as a retailers' margin.

The Clarion - *Summer 2007*

Seasons and Cropping

Seasons of the Year

Farming since time immemorial has depended on the seasons of the year, and the farming calendar has to work with those seasons. In the current month of June we will soon see the longest day of the year, in July the oilseed rape and barley harvest will start and in August most of the wheat crops will be harvested.

Wheat Cropping

Wheat is the most important crop on arable farms (the dictionary definition of arable is "fit for the plough; suited to the purpose of cultivation") since the highest quality wheat grains are milled to make the flour which is the main ingredient of bread and biscuits. Wheat is also used in some breakfast cereals, whilst the lower protein wheat is used in most poultry and livestock feeds.

Many of the cropped fields in our Clarion area are winter wheat, seeded in September or October to produce a winter hardy plant which is then ready to grow fast when the soil temperature and length of daylight increase in the Spring. At the present time of year, in June, the wheat fields should be a dense stand of dark green tillers, with the seed heads which have formed in the stems starting to appear above the flag leaves.

During the Spring and early Summer these wheat crops will have been inspected frequently by the farmer, often with an Agronomist, to prescribe the best crop protection treatments to ensure healthy growth. These treatments will have included plant nutrients for growth, herbicides to kill weeds, and fungicides to protect against fungus diseases such as mildew. Some crops may have been sprayed with growth regulator (to shorten and thereby strengthen the stems) or with insecticide if there has been a problem with aphids.

In July the wheat will start to turn in colour from the green of the growing crop to the golden colour of harvest, and the combine harvester will start its work when the grain is hard with low moisture content, and hopefully the sun is shining.

Public Footpath in crop of oil seed rape.

Oilseed Rape and Barley

A few local fields are cropped with oilseed rape grown for the vegetable oil produced by crushing its small black seeds. This most distinctive crop, with its bright yellow flowers in April, is normally harvested in July, for use as cooking oil or for the manufacture of margarine.

Other arable fields are growing barley, which is a paler shade of green than wheat and will have come into ear much earlier. Its spiky awns first appear in May, and the crop should be ready for harvest in July. Most average quality barley will be used for pig and cattle feeds, but the best quality barley will be used for malting to become the main ingredient for brewing beer, with hops added to give the beer its traditional flavour.

The Clarion - *Autumn 2007*

The Farmer's Share of the Loaf of Bread

The Guernsey dairy herd walking out to pasture after milking at Lacey's Family Farm, Bolter End.

Prospects for the coming (2008) Farming Year

This is the time of year when the cereal harvest has been completed and farmers measure up the quantity of grain in store and appraise its quality and value and ponder on their prospects and the effect that weather has had on the past year's work.

Climate Warming

Climatologists warn us that global warming will lead to more "extreme weather events". This certainly seems to have happened over the past year in which we experienced a very wet October (136mm of rainfall) causing some fields to cap after seeding, thus making it difficult for the new seedlings to poke through this hard surface layer of soil. This might not have mattered if the crops had experienced good growing conditions in March and April, but instead we had the driest April on record, with just 3mm of rainfall. Then we moved into the Summer season, when everyone expects plenty of sunshine, both for themselves and for the crops, but Summer 2007 produced 400mm (about 16 inches) of rain in the three-month period from May to July, and with it the floods which were so devastating especially near the River Severn, but also caused flooding in the River Thames valley so that livestock had to be moved to higher ground. Fortunately there were good days at the beginning and end of the harvest month of August, but overall cereal yields this year have been at least 15% less than in a more normal year.

The Farmer's Share of the price of a Loaf of Bread

So quantity of grain is down as a result of the weather in Britain, and overseas drought in Australia and other parts of the world has resulted in lower yields, but demand for grain has risen. This has been due to increasing world population, and the higher standard of living and eating in China (following its success in commerce and manufacturing industry) and also the quantity of maize and other crops being used in USA and Brazil to produce bioethanol as a fuel. This change in the ratio of supply compared to demand for grain has resulted in an increase in the value of grain, so that now the farmer's share for the wheat in a loaf of bread has risen from 8 pence last year to 12 pence this Autumn.

The Great Milk Debate

But we must not forget the livestock farms, since an increase in the value of grain means that animal feeds have become more expensive, and the stock farms have also had the worry of the recent outbreak of foot and mouth disease which appears to have started in the Government research station at Pirbright in Surrey. However the joint WI and NFU initiative in May "The Great Milk Debate" was a real encouragement to dairy farmers, by proving how much their milk is appreciated by consumers and by the consequent fairer share of the supermarket retail price of milk which is now allocated to the farmer as a result of the Women's Institute campaign.

The Clarion - *Winter 2007*

The Urge of Plants and Animals to Reproduce

We are approaching the middle of winter, and those gardeners who have a lawn that must be mown regularly in the Summer to keep it at its best have probably cleaned their mowers, wiped them with an oily rag and put them away in the shed - where most of them will rest inactive until the soil temperature warms up and the days get longer next Spring.

Leaves and Seed heads of Grasses

It is interesting to compare lawns on the garden scale with pasture on the farm scale. Lawn grasses, such as chewings fescue or bentgrass will have narrow leaves and will thrive in a dense short sward cut regularly with the lawn mower. Pasture grasses on the farm, such as perennial ryegrass or timothy will have a wider, taller and more succulent leaf which will benefit from regular grazing by cattle or sheep, and this will be the main feed for the cattle and sheep throughout the Summer grazing season.

Why does the grass grow again in this way after each cut of the lawn mower or grazing by the cattle and sheep? The answer is that the strongest urge of most plants and animals is to reproduce - to ensure that there is a next generation. In the case of grass plants this means to form a seed head. So to keep a lawn or a pasture in good order grasses must not be allowed to go to seed, and then they will grow again vigorously in another attempt to form a seed head - and they will continue to re-grow after each cut until the days get shorter and colder and it is no longer the right time of year to form seeds or to have sunshine to ripen them.

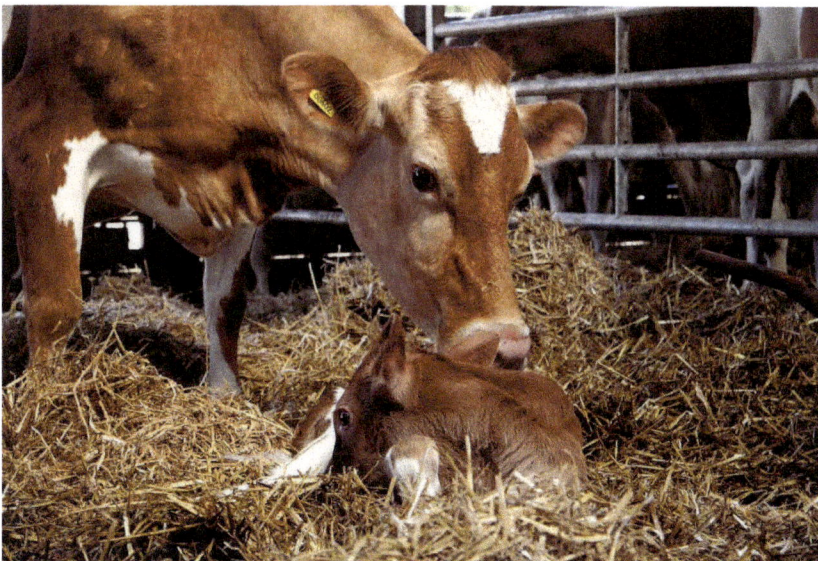

The cow, before selective breeding, would only have produced sufficient milk to feed its own calf.

Why does the Dairy Cow produce Milk ?

On the farms during the Winter the cattle and sheep have to continue to eat and grow, and in the case of dairy cattle to produce milk - and we must remember that the dairy cow really only produces the milk to feed its own calf, and that the milk which we humans buy from the dairy or shop is just the extra milk produced by the cow (after hundreds of years of selective breeding) which is too much for the baby calf to drink.

So while the grass is not growing in the Winter the livestock on the farms are often fed with hay, made from grass which has been allowed to grow up and form seedheads and then cut, dried and made into bales during the hot days in the Summer. Silage is another good method of preserving some of the Summer grass for use in the next Winter. One way of making silage is to cut the lush grass, often in May, then after allowing it to wilt it is moved to a 'silage clamp' where it is compressed by rolling it with a tractor so that anaerobic bacteria form lactic acid in a controlled fermentation. This lactic acid preserves the grass in just the same way that small onions can be put into a jar of vinegar (which is an acid) to make pickled onions.

The Clarion - *Spring 2008*

Good Seedbeds. Blue Tongue of Cattle

Weather conditions at Seeding Time

In the September issue of The Clarion we looked at global warming, the wet summer of 2007, flooding, world supplies of food and the price of bread and milk. What has happened in the last six months?

Firstly the good news, that last September and October were good months for weather with 81 mm (just over 3 inches) of rain in the two months, and with plenty of bright sunny days, and this resulted in ideal soil conditions on all the fields which were being planted with autumn sown wheat ready for harvest in August 2008.

Importance of a Good Seedbed

Despite all the modern advances in the design of tractors, such as the size of the 4 wheel drive tractors which have been ploughing the fields near Lane End last September developing 275 horsepower from their 8.1 litre engines, and their sat-nav monitors in the cab to measure the size of the field and to steer the tractor along a line precisely parallel to the previous pass across the field and the right distance away from it, good weather and soil conditions are still the most important factors to make a good seedbed. And a good seedbed is just as important on the farm as in the vegetable garden to start a healthy and high yielding crop. So the news on the arable farms, where the milling wheat for bread making is grown, is good news with the crops looking well.

'Blue Tongue' disease of Cattle

But the livestock farms continue to have difficulties, both with their production costs increasing more than the farmgate price of the food which they produce, and with disease adversely affecting the health of their animals. There is a disease new to Britain called 'Blue Tongue' which is likely to become very serious amongst cattle and sheep when the warmer weather comes. This disease is spread by infected midges which bite the cattle and sheep to suck their blood, but in so doing they infect them with the blue tongue disease. This disease used to be only seen in Africa where the hot climate suited the midges, but last year it spread to Belgium where 90% of the cattle became infected and 11% to 15% of them actually died. Then on 5th August 2007, when the wind blew from Belgium across the Channel taking the midges with it, the cattle and sheep in Norfolk became infected. The English Institute for Animal Health have developed an inactivated vaccine to protect the cattle and sheep, and they hope that 22 million doses of it will be available by May 2008. But they do not know if this will be sufficient or if it will be soon enough.

The Clarion - *Summer 2008*

Silage. GM Crops

Nigel Rogers, in the 1960's, using the Self-filling Silage Trailer designed by the author and built by Eric Meakes at the Blacksmith's Forge at Lane End. The author patented this design, but it was not taken up by any manufacturer. However this prototype successfully harvested all the Kensham Farms' silage for eleven years until the dairy herd was sold in 1970.

Silage = Pickled Grass

I write these thoughts during an exceptionally wet Spring Bank Holiday, not the right weather at all for folk who are camping. But for those of us who grow wheat the weather this season so far has been favourable, with a strong plant established in good soil conditions last September and October followed by sufficient rainfall and good sunny days to keep the crop growing. So at this stage the crops, as well as lawns in private gardens and the foliage on woodland trees, are all looking luxuriant and healthy.

But the livestock farmers are trying to make silage on which to feed their livestock next winter, and for that job fine weather is needed. Silage is pickled grass, preserved in naturally formed lactic acid when conditions are right, but this natural fermentation goes wrong if the grass is too wet and then foul smelling butyric acid would be formed and the silage would be unpalatable and of low nutritional value. Next winter may seem a long way ahead, but the fodder which the cattle will eat during the winter months, whether it is silage or hay, all has to be grown during the Summer, especially in the months of May and June.

GM – Genetically Modified Crops

I will now mention the controversial subject of genetically modified (GM) crops. Some folk regard GM foods as being too unnatural and a health risk, but 500 million hectares of GM crops are grown worldwide and no personal injury or ill health nor any disruption of ecosystems has ever been detected from their growing or use. Benefits can result in higher crop yields with low inputs of crop protection chemicals. Most of the foods we eat, such as bread, milk and meat have been modified by selective breeding and it is only a few hunter gatherer foods such as fish and wild mushrooms or berries that have not been made more plentiful by selective breeding.

Genetically Modified crops are developed by taking genes from one organism, such as the genes in peas or beans or clover that enable the plant to use atmospheric nitrogen as a fertilizer, and then transferring those desirable genes by means of an agrobacterium into another type of plant such as wheat. Thus GM wheat which could grow well without using a lot of nitrogen fertilizer could be developed by using this new technology. Other examples are the American varieties of soya beans and maize which have already been developed with GM technology to be immune to certain weed killers, so that weeds in those growing crops of soya or maize can be effectively controlled by using that specific weed killer.

UK Government does not allow the growing of GM foods in Britain at the present time but there are now worries about the world supply of food. It seems unfortunate that so much basic food research, such as GM technology, is now carried out by multinational corporations with vested interests, rather than by impartial Government research stations or University departments.

The Clarion - *Autumn 2008*

Bushel Weight. Gluten Content of Wheat. Drying Grain

Cropping on the Chiltern Hills

What has all the wet weather in August done to the crops ready for harvesting? This question deserves a serious reply, because the rainfall in the Spring and Summer 2008 which helped all the crops to grow so well from April to July then carried on during August to make harvesting really difficult.

In the Chiltern Hills potatoes and vegetables are not often grown as farm crops since the flints in the soil, the 600 feet altitude above sea level and the soil type all make the local fields more suitable for growing either cattle and sheep feed, such as grass or maize for silage, or alternatively 'combinable crops' such as oilseed rape or cereal crops, of which wheat for making bread, biscuits and breakfast cereals is the main crop.

Bushel Weight

Wheat that is suitable for making bread will have been grown from seed bred by Plant Breeders especially for its high protein, ideally around 13% protein, and good 'bushel weight'. The 'bushel' was an old imperial measure of 8 gallons, used for grain and other animal feeds. A bushel of good quality wheat weighed more than the same measure of poor quality wheat containing shriveled grains and husk. In modern metric terms the good sample of milling wheat might weigh 76 kilograms per hectoliter, whereas the poor quality sample, only suitable for animal feeds, might weigh 69 kg/hl.

Gluten Content of Wheat – the Hagberg Falling Number

The colour of the crops at harvest time is also important, the ideal being a golden yellow colour shining in the sunlight of the harvest field. But this year the excess rain on crops that had already ripened caused them to change to a grey colour, with a risk of moulds or mycotoxins forming on the straw and the seed head. We have had a struggle to harvest these grey looking crops on days when the sky was equally grey. But the worst part of this damage to the ripe grain is that the gluten content is reduced, measured by a clever test to show the 'Hagberg Falling Number'. When the gluten quality is reduced by rain at harvest time in this way the loaf of bread will not rise properly when it is baked, and in extreme cases might result in a stodgy loaf. If the quality of English wheat is too low due to the wet harvest, then the millers and bakeries might have to import more wheat from countries such as North America with a drier summer climate.

Wheat ready to harvest.

Drying Grain

We all know that fresh lawn mowings will heat up on the compost heap, and then rot down. The same process would happen to cereal grains if they were to be stored when damp. So in order to store the grain well, sometimes for as long as eight or ten months, the grain will have to be dry and cool. In a year like 2008 this will mean that the grain dryers, running on expensive oil and electricity, will be running for longer than in those years when the grain had been harvested dry on bright sunny days.

The Clarion - *Spring 2009*

Vernalisation of Winter Wheat

Spring seeding of wheat with the Weaving seed drill. In the background the Class crawler tractor is rolling in, following seeding.

Spring Seeding of Wheat

It was a real encouragement to start seeding spring wheat on 23rd February, into seed beds that were sufficiently dry to take the weight of the tractor and seed drill. The past season has been quite a reminder that farming has to work with the weather, since the excessively wet harvest month of August 2008 was then followed by so much Autumn rainfall that not all of the planned winter wheat sowings could be carried out, although we were able to seed the winter milling wheat contracted for sale in Autumn 2009 for Warburtons bread.

Vernalisation in the Cold Winter Season

Varieties of wheat that are suitable for seeding in the Autumn months of September to November need a period of hard weather for vernalisation, in which the young plants recognize that winter is over. If these winter varieties of wheat were to be seeded after mid-February the young plants would not be subjected to winter temperatures, and so they might only grow leaf in the first season without flowering and forming seedheads until the following season.

There are many fields have been too wet for seeding in Autumn 2008 so that they will now have to be seeded with spring varieties of cereals which have been selectively bred to flower and form seedheads in the same season, without need of any period of cold weather for vernalisation. These spring varieties have a short growing season, coming to harvest around five months after seeding.

In practical terms we have found ourselves at Kensham farm in the same situation as many of the arable farms in England. Due to the wet Autumn about a quarter of the winter wheat seed which we ordered could not be sown in time for this cold weather vernalisation. This seed will have to remain in the store for a whole year, to be sown in Autumn 2009, providing that a germination test shows that the seed to be still viable at that time. Meanwhile we have had to buy additional seed corn of spring varieties for seeding in February and March for the fields which could not be seeded in the Autumn. This will be spring barley, which we hope will be suitable for making the malt for beer after it is harvested in August. We had already planned to grow some Canadian hard red spring wheat, a high quality but low yielding milling wheat which is used for making Hovis bread.

Asbestos Cement (Chrysotile) Roofs on Farm Buildings

Asbestos cement barn roof. The section covered by moss was over 50 years old. Roof repair needed only a few replacement sheets. Crysotile ("White Asbestos") roofing sheets have never been shown to have caused any measurable risk to health.

Barn Roofs – Corrugated Chrysotile Asbestos Cement

We are now in the middle of the growing season so that on the livestock farms the cattle and sheep will be out in the pastures grazing grass as it grows. Other grass fields will have been specially fertilized and then shut up for making silage or hay to be stored in farm buildings until needed for consumption by livestock next Winter. And farm barns must have roofs, preferably with gutters, to store this hay and fodder in good condition and there must be yards where the cattle can live in the Winter, and grain stores where wheat for human consumption can be dried and stored until required by the millers and bakeries for preparation of bread, biscuits or breakfast cereals.

In years gone by many of the roofs of farm barns were made from local materials, such as thatch from straw or reeds grown nearby or from roofing tiles like those which used to be made from the clay quarried locally from Ditchfield and Cadmore End Commons and baked in adjoining kilns fueled by firewood from the same commons. But most of the barns and farm buildings built since World War II have been built with roofs made of corrugated asbestos cement, an exceedingly safe and practical material which has been erroneously classified as 'hazardous' by our less than competent masters in the EU headquarters at Brussels.

The British Health & Safety Executive Research paper HSL/06/55 on Risk Assessment of Materials containing Asbestos

Many farmers are now pressing for recognition by the EU and British Government of the safety and advantages of asbestos cement sheets in which Chrysotile, generally known as 'white asbestos', is the natural magnesium silicate fibrous rock mineral, mined mainly in Russia and Canada, used as reinforcement to give strength to the cement and aggregate component of the sheets. In 170 other countries of the world, including USA, Canada, Brazil, Russia, India and China asbestos cement roofing sheets continue to be manufactured and used without any measurable risk to health.

The EU legislators have unfortunately confused the safe white asbestos (Chrysotile) with the dangerous blue and brown forms of asbestos (Amphiboles) which were used many years ago for lagging pipework in ships and old-fashioned boilers. The blue and brown asbestos was a completely different acid resistant mineral with sharp needle like fibers which could be breathed into the lungs, and being acid resistant remained in the lungs, sometimes causing illness or death many years later.

However, regulation of the asbestos industry in Britain started in 1931, and by 1979 Britain's Health & Safety Commission had imposed a complete ban on the use of the dangerous blue asbestos. In 2006 the HSC published a research paper (HSC/06/55) showing that work with asbestos cement does not involve any significant risk to health, and that it is considerably less dangerous than normal work in the construction and agricultural industries.

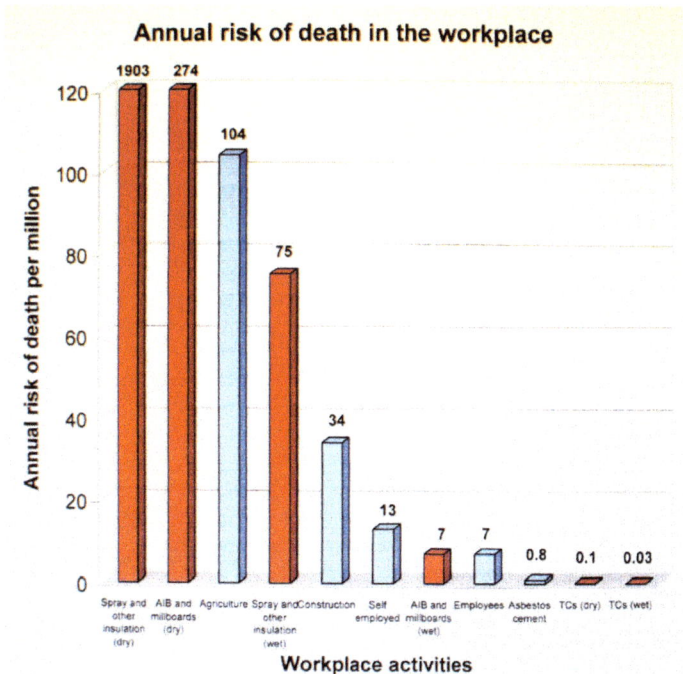

Annual risk of death in the workplace

This is Figure 3 from the Health and Safety Commission research paper HSC/06/55. It will be seen that the risk of death from working with Spray and Other Insulation (dry) at 1,903 deaths per million workers is so great that only a fraction of the risk can be shown on the appropriate bar above, which is only calibrated up to 120 deaths per million. Of the other work place activities, it will been seen that agriculture at 104 deaths per million is more dangerous than most other work place activities. However the risk from work with asbestos cement is less than 1 worker per million per annum.

Wheat quality. Bushel weight. Protein. Fungicides

Cultivation and Preparation of Seedbeds

Harvest 2009 is now over, and September is always a good month in which to appraise the last season. How did it all turn out? And how can we improve our cultivations or timings for next year?

Yields of winter wheat from all of the fields that were seeded in September or October 2008 have been disappointing, largely because the seedbeds were damp and poor, and then it rained at the wrong times. Farm fields are the same as vegetable gardens in that for best results the seeds should be sown into seedbeds which are so dry and friable that the gardener can wear shoes rather than boots, and the tractor can drive over the fields without leaving ruts. Then, in an ideal season, there will be plenty of rain while the young plants are growing fast in the Spring and early Summer, and this will be followed by drier weather as the crop ripens. Unfortunately, in 2009 the months from March to June were all unusually dry, and then July was excessively wet. But this was followed by two dry weeks in the middle of August which were just right for harvesting most of the cereal crops.

Wheat Quality

Quality of the wheat harvested this year has been good with excellent bushel weights, since the individual grains are plump and heavy. The protein readings are high and the gluten content satisfactory, so most of this wheat from milling varieties should reach the high standards required by the millers and bakeries for bread making. Some of the spring barley, seeded in February or March, has also grown well and should reach the malting standard required for brewing beer. Protein levels in grain are interesting, in that the best milling wheat will have high protein whereas the best malting barley will have low protein.

CYO mobile seed cleaner, with which farmers can save their own grain for use as seed for the next season. This cleaner removes any impurities and dresses the seed with fungicide.

Protecting the Health of Growing Crops

In the vegetable garden diseases such as club root in cabbages can sometimes be a problem. Similarly, farmers have to take precautions to protect the health of the cereal plant throughout the growing season. Before the seed grain is sown it will have been dressed with a seed dressing powder to give protection against diseases such as bunt and loose smut, and the newer seed dressings also incorporate a systemic insecticide which will protect the young seedlings for the first six weeks of their life from attack by aphids.

Later in the life of the cereal plant spray treatments with fungicide will be necessary to protect the leaf against diseases such as powdery mildew, septoria and yellow rust. But there is one disease of the root of the cereal plant called 'take-all' which has been troublesome this year, following the wet and unsatisfactory seedbeds in Autumn 2008, for which no satisfactory plant protection chemical has yet been developed. The symptoms of take-all are a black fungus on the root system followed by pale bleached looking straw carrying empty seed heads. But it has been found that when cereal crops are grown continuously in a field it can develop a natural immunity called 'take-all decline'. This makes it possible for cereals to be grown continuously in the same field without using crop rotations.

The Clarion - *Winter 2009*

Preparation of Seedbeds. Minimal Cultivations versus Ploughing. Defra Statistics on Imported Goods

Autumn ploughing with an 8 furrow reversible plough.

Autumn 2009, in particular October, will be remembered as a pleasant dry and easy month – just right for the arable farms where work on ploughing, cultivations and seeding ready for harvest 2010 took place in ideal conditions. Ploughing and cultivations started straight after the combine harvester had finished each field.

Preparation of Seedbeds
The preparation of seedbeds on the farms is the same in objective, but different in scale, to work in a vegetable garden. Germination of the new seeds will be disappointing if there are hard clods of earth with cavities in between. On some farms with heavy clay soils the land was ploughed to bury the stubble and trash and then not worked for two or three weeks. On those clay soils some of the clods dried out to become as hard as bricks, and nearly the same size as a half brick, whereas if the ploughed land had been worked with disc harrows or cultivator before it became too dry then a fine and friable seedbed could have been achieved.

Minimal Cultivations versus Ploughing

Some farms use 'minimal cultivations', where the first implement on the stubble will loosen and stir the soil rather than turning it over. This can only be successful if the stubble is clean and free of weeds, which could subsequently re-grow and choke the next crop.

At Kensham Farm we use the plough on some fields and minimal cultivations on others, depending on the soil type and how clean the stubbles are after harvest. But with both methods we try to make the seedbed fine and firm to discourage the movement of slugs, and to make it soon after harvest to give the weed seeds time to chit and start growing. The very young weed plants can then be killed out with a herbicide spray such as Roundup before seeding the wheat or barley in September or October, ready to be harvested during the next August.

Defra Statistics on Imported Foods

On the political side of farming and food production we farmers are pleased that Defra, the Department for the Environment, Food and Rural Affairs which is responsible to the EU in Brussels for administration of the many regulations affecting British farms, has recently published statistics on food supplies. These statistics show that fifteen years ago we grew 77% of the fresh vegetables that were consumed in Britain, whereas now we only grow 50%, and the proportion of indigenous foods, that is foods which can be grown in the British climate, has dropped from 82% down to 73%.

This comes as no surprise to us farmers, since politicians of both major political parties for the past twenty or more years have favoured imports of cheap food from overseas countries with lower quality standards than our own. However, it is really good that Government is now recognising the importance of food security, and we hope that Defra's future policies will encourage British farms to grow more of our own food.

The Clarion - *Spring 2010*

Effect of Frost and Snow on Cereal Crops. Plant Nutrients. Campaign for the Farmed Environment. Sewage Sludge.

Effect of Frost and Snow on Cereal Crops

Many neighbours have asked me how the crops fared in the recent hard winter when heavy snow and freezing conditions started on 18th December and continued with some roads closed and other interruptions of the daily routine of life until mid-January. But this did not harm any of the winter wheat or other crops, which had all been seeded in good dry conditions in September or October 2009. The crops were really quite snug under the snow and looked good after the snow had melted. However, if the weather had been dry with bitingly cold winds from the north east for such a long period then the crops could have suffered damage from wind chill.

As we approach the Spring we have to ensure that the young crops have sufficient plant nutrients to draw on for good growth. When soil temperatures rise and the length of daylight increases it is a signal to the young plants that the dormant winter period has ended and the main growing period of the year from April to June has started.

Plant Nutrients

These main plant nutrients are often shown on a bag of fertilizer as N, P and K. These letters are abbreviations for Nitrogen for leaf growth, Phosphorous for a vigorous healthy and fibrous root system and Potash (denoted by the letter K) for general plant health and the production of starch and sugars in the leaf as it grows. All of these nutrients only work well when the soil is at the correct pH level; if the soil is too acid this can be corrected by adding lime.

Another nutrient which the plant needs, especially in crops such as oil seed rape - the fields which are bright yellow in early Summer - is Sulphur. It is interesting that when there was more manufacturing industry in Britain there used to be sufficient sulphur in the air for the growing crops, but following the decline of industry there has been less air pollution so that now we have to buy fertilizer that contains sulphur.

Campaign for the Farmed Environment

At a recent farmers meeting in Oxfordshire the Chairman of the Government Agency 'Natural England' introduced the latest Defra initiative, the 'Campaign for the Farmed Environment' to encourage Environmental Stewardship measures aimed to protect the countryside, its wildlife, soils and water quality. At that meeting we were reminded of the alarming fact that there are only sufficient natural resources of Phosphorous in the whole world to last for the next sixty years. After that, future generations will have to grow food without the present resources of phosphate fertilizer, and also without the present natural resources of oil both to power tractors, and as the base ingredient of many crop protection chemicals.

Sewage sludge supplied by Thames Water being loaded from a heap in the corner of the field onto the contractor's 5 wheel spreader.

Sewage Sludge

On the farm we use large amounts of artificial fertilizer, but we also aim to use sufficient digested sewage sludge, processed at the sewage works near Twickenham which serves the heavily populated London suburbs, to maintain the Phosphate status of the soil at the best level. We use a dressing of eight tonnes per acre of this treated sewage sludge every second or third year on many of our fields. Gardeners on their vegetable plots are similarly well advised to make and use compost, both to return plant nutrients to the soil, and also to add humus to keep the soil open and make it easier to work.

The Clarion - *Summer 2010*

Volcanic Dust grounding Air Traffic. Badgers infected with Tuberculosis. Protection of Badgers Act 1992. Open Farm Sunday.

Volcanic Dust grounding Air Traffic

Since our last issue of Clarion V in March we have had a week of disruption of air traffic due to volcanic dust from Iceland, with all air freight and passenger services in Britain grounded for several days. In the same week we have watched the leaders of the main three political parties in debate, from the same platform in television studios, explaining their intended policies for a new Government to follow the General Election.

This is all quite relevant to our farms and food production. Two years ago Professor Patrick Wall, Chairman of the European Food Safety Authority accused EU politicians of "over-reacting to small risks" and of putting politics before science. Was the grounding of all aircraft in April an over-reaction to a small risk or was it good science and a real risk?

Badgers infected with Tuberculosis

Similarly are the laws that protect the lives of diseased badgers sensible and based on good science at the present time when more than 40,000 cattle in England and Wales are being compulsorily slaughtered each year by Ministry officials, because they have failed the Tuberculosis Test? Forty years ago, before the badger protection laws were passed, tuberculosis was a disease which had been almost eliminated from both the human and cattle populations of Britain. Vets can now analyse the DNA of mycobacteria taken from the lungs of badgers killed by road traffic and find that this corresponds to the same sub-type of mycobacteria found in diseased cattle within the same parish or area. Thus DNA testing, similar to the tests used by the Police to solve crimes, can also prove the connection between infected wild badgers and cattle infected with bovine tuberculosis on farms in the same area.

Last Autumn a retired farmer friend on the border between North Devon and Cornwall recounted to me the sad tale of the farming family who had bought his farm, less than a mile from the smallholding where he now lives. This farmer's dairy herd had become infected with bovine tuberculosis, almost certainly from infected wild badgers which live in the setts near to his farm. This resulted firstly in the Ministry officials from Defra imposing movement restrictions on this Devon farmer whereby he was not allowed to move any cattle on or off his farm, and then secondly his licence to sell milk was withdrawn. So his cows had to be slaughtered, he and his family were in emotional turmoil, as well as being ruined financially, and no one was allowed to put the diseased badgers out of their misery as they died a slow and painful death.

Badger sett in the middle of a field of wheat. The damage to crops is trivial when compared to the risk to cattle of bovine tuberculosis transmitted by infected badgers.

Protection of Badgers Act 1992

A further most unfortunate outcome of the Protection of Badgers Act 1992 is the destruction of the nests of ground nesting birds such as lapwings, partridges, larks and willow warblers by badgers, and the number of hedgehogs which are now being killed by badgers. Inept Government interference with the work of naturalists, farmers, landowners and their gamekeepers has so favoured badgers that they have greatly increased in numbers, some healthy but others suffering from TB, to the detriment of these other wild animals and birds as well as dairy and livestock farming. So the best solution would be a repeal or modification of this Act, and less Government interference in the future.

Fortunately on the Chiltern Hills there does not seem to be any disease amongst the badger population yet, even though there are probably three times as many badger setts in the woods and fields nearby than when we started farming at Kensham Farm fifty five years ago. The only dairy herd remaining in Clarion V territory, Laceys Family Farm at Bolter End, is thriving with clear tests for tuberculosis and providing excellent quality milk and eggs from the family's farm shop.

Open Farm Sunday

Open Farm Sunday this year will be on Sunday 13th June 2010, when more than 400 farms will be open to the public to show and explain the work of growing crops and producing food. Here at Kensham Farm, Cadmore End, HP14 3PR we will be open from 2:30 pm to 5:00 pm and hope to welcome some of our Clarion V readers. More information, with a list of all the other farms that will be open may be found on < www.farmsunday.org>

The Clarion - *Autumn 2010*

Saddle Gall Midge. Pond Restoration. Food Security

Plant Deformities caused by Insects

During the long hot early Summer of 2010 we have been troubled by two particular insect pests on our wheat crops, both of which live in cocoons in the soil from year to year, and then only become troublesome when soil temperatures warm up in May to mid-June.

Vegetable gardeners always look out for aphids on their rows of broad beans, and similarly cereal farmers always look out for the Orange Blossom Midge which emerges when soil temperatures reach about 13°c. The emerging midges lay their eggs on the floret of the wheat ear as it is forming, and the eggs soon hatch into larvae which spoil both the quality and the yield of the grain as it forms in the seed head ears of the wheat. The remedy when orange blossom midges are flying around the young wheat crop in May is to spray them with an insecticide which will kill the flying midges, or with a stronger insecticide which will be effective for a week or ten days against the young larvae feeding on the wheat as well as against the flying midges.

Saddle Gall Midge

This Summer many wheat crops have also been affected by the most unusual Saddle Gall Midge. This has a similar life cycle to the orange blossom midge, but the midge does not hatch out from the dormant cocoons in the soil until soil temperatures have reached 18°c. The flying midge then lays its eggs on the leaf sheath of the growing wheat plant, and these eggs hatch out into larvae, which are bright red or orange grubs about 5 millimeters in length. These small grubs then move down the leaf and damage the stem of the cereal plant by feeding on it. This year was the first year in which we had been troubled by the saddle gall midge, since it is generally only a problem in warmer climates, and by the time we had spotted them it was too late to take any protective action. After harvest we will know how much yields have been depressed by the stems that have transverse swellings caused by the saddle gall midge.

By early July the poor growth on pasture fields caused by lack of summer rainfall had become a serious matter for dairy and livestock farms, to the extent that some farmers were having to supplement the grass grazed by their cattle with hay or silage reserved for the coming winter.

Stem of growing wheat plant infected by saddle gall midge larvae.

Pond Restoration

A side effect of this lack of rainfall for us was that the pond on Cadmore End Common in front of Kensham Farm totally dried out, the first time since 1975. By the end of July there was so little water in the pond that we had to provide water in a pan for the newly hatched moorhen chicks. We walked over the edges of the pond and found that there was of depth of more than two feet of soft dried out silt above the original hard clay bottom of the pond. This pond is not fed by any spring but instead it collects surface water from several acres of fields as well as from the gutters, yards and roadway around the farm buildings, so the deeper it is for collecting of water in a wet period then the longer will it retain water when evaporation takes place in a long dry summer.

So we decided to act immediately, before the next rainstorm which might refill the pond, and to clear out the soft silt using our Manitou rough terrain fork lift with large 2 tonne capacity bucket on its telescopic arm. We used this to scoop out the silt, together with goat willow bushes growing in the silt, to load onto grain trailers and tip onto a heap for spreading on the fields at a later stage. In this way we moved about 350 tonnes of silt and bushes within two days, and then tided up and re-seeded the pond verges where the farm machinery had left tracks.

Pond cleaning is just one of the many conservation tasks that we try to fit in with the farm's real work of growing wheat for bread making. Government and EU policy in recent years have been encouraging conservation work with grants towards the cost of the work, but in the case of cleaning the Kensham pond it seemed more sensible to carry out the work straightaway while weather conditions were right, rather than to spend months on the correspondence and preparation of grant application forms that are necessary before grant aid can be approved.

Food Security

The recent good news on the political front is that our Coalition Government recognizes the importance of an adequate supply of food for the people of Britain. For the past twenty years Government policy favoured importing food from overseas if it was cheaper, and failed to recognize that if there were ever to be a worldwide shortage of food, then the countries abroad would stop exporting their food to us in preference to letting their own people go hungry. But now we have a Government that realizes the importance of food production from British farms.

The Clarion - *Winter 2010*

Excess of Rainfall in August. Ventilated Floor Grain Dryers. Continuous Flow Grain Dryers. Russian Grain Harvest. Wheat Exports to Morocco.

Excess Rainfall

In the last issue of Clarion V the summer drought was mentioned, with an average rainfall of less than 30 millimeters in each of the growing months of April, May, June and July - but at that stage we had not realized that in the harvesting month of August, when ideally the weather is hot and sunny while farmers are harvesting and other folk are away on holiday, we had 124 mm of rainfall - more rain than the total in the four growing months. So that made harvesting difficult, with many of the days so wet that the combine harvester could not be used, since it would have become clogged by wet straw wrapping around the thrashing mechanism.

But the aim in a wet and difficult harvest is to get the grain harvested and put in the grain store whether or not it is sufficiently dry to store. Once the grain is under a roof it will be safe, but it must then be dried down to a moisture content of less than 15% in order to store safely and to be saleable.

Ventilated Floor Grain Dryers

At Kensham Farm we use ventilated floor dryers, in which there is a cavity below the floor, covered with alternate hardwood and perforated metal strips. The perforations are of a size that allows airflow without being so large as to allow the grain to fall through. The grain trailers are driven over these floors before tipping and the grain is piled up to a height of 8 feet or even 10 feet. The drying fans in the air tunnels connected to the floors of the grain store cavities are about 3 feet in diameter, each powered by a 25 horsepower motor, with a second fan boosting the air flow of the first fan so that the dry air will be forced upwards through the grain.

This type of grain dryer works on the same principle as drying laundry on the washing line on a bright windy day - it is the flow of dry air that does the drying. On the brightest days there is no need for any additional heat, although on the duller damp days a Calor gas heater provides enough heat to dry the air, without significantly heating it. But it is a slow process which can take several weeks to bring very wet grain of 20% or 25% moisture content down to the level of 14% or 15% at which it will be saleable and safe to store.

Continuous Flow Grain Dryers

At Myze Farm, on the A40 road between West Wycombe and Piddington, there is a "continuous flow" dryer. That is a different process, dependent on an oil-fired furnace with a fan which forces hot air through a sloping tray of moving grain. This is more like a cooker than a washing line, but it is the best method if the grain comes in from the harvest field with excessively high moisture content.

Alvin Blanche continuous flow grain dryer with capacity of 10 tonnes per hour at Myze Farm.

Russian Grain Harvest

In financial terms the low yields and wet difficult conditions of the 2010 Harvest have not been a disaster, since the harvest in Russia was so disappointing that exports from Russia to other countries have been stopped for a two-year period. This has led to fear of shortage on the world grain market, and the marketing principles of supply and demand has caused the value of grain to increase. The value of the milling wheat in a standard loaf of bread has increased from about 9 pence in Spring 2010 to 14 pence this Autumn, meaning that loaf of bread at the Supermarket should not have increased in price by more than 5 pence.

Wheat Exports to Morocco

Fortunately, the breadmaking quality of the wheat from the English 2010 harvest has been better than in the rest of Europe. In the first week in November 1,000 tonnes of milling wheat was collected from Kensham and Myze Farms, 35 lorry loads of 29 tonnes each, for export from Portbury Dock near Bristol to go for breadmaking in Morocco, but some of the remaining 4,000 tonnes from the August 2010 harvest may not be collected from the farm until May or June 2011.

Sustainable farming practices in the Chiltern Hills

British Agriculture - Is It Sustainable?

My grandson, at his Agricultural College, has been set the subject of an interesting essay to write on *"What contribution does farming practice in the Chiltern Hills make to British Agriculture - and is it sustainable?"*. I must not write his essay for him, but I will set out some of my thoughts on the sustainability of arable farming in this area.

Economic Sustainability

This is the important part of any self-employed activity or farm business. Will there be a profit? If there is no profit in a complete season then the farm business will not be sustainable, since the farmer's family and living expenses have to come out of the profit. No profit would thus result in no wages and no money with which to run the farm business.

The tools for economic sustainability are budgets and forecasts - if there is a good business plan on paper then in practice it may or may not work out well. However, if the forecast plan shows a loss on paper then in practice it cannot succeed, meaning that such production would not be sustainable, and a different business plan for stocking or cropping on the farm will have to be thought out and put into practice.

Agricultural Sustainability

Those who have studied agricultural history will have heard of the 'Dust Bowls' in North America from Kansas down to Texas in the 1930's. There, on the arable land of the Great Plains, farming practice had been bad, in that humus (i.e. the organic matter) in the soil was not replenished, and there were no windbreaks such as trees or hedges. Hot summers, drought and strong winds caused the land to become like a desert, where the young growing crops were blown away and no food could be grown.

When growing cereal crops, such as wheat for breakfast cereals or for milling into flour to make bread or biscuits on the Chiltern Hills near Lane End we have to be certain that we maintain the soil in good condition. We ensure the seasonal requirements of this year's crop by testing the soil for acidity, and then correcting fields which are too acid by spreading lime in the form of ground chalk. We also test the soil for phosphate and potash levels, and correct any deficiency with fertilizer out of a bag, often supplemented with farmyard manure or sewage sludge - both of which are rich sources of humus that will maintain the soil in good condition so that it is easy to work and fertile, and will not lose its structure in the way that happened in the American dustbowls. Nitrogen will also be required for leaf growth, and then when a strong healthy plant is grown the significant residues after harvest of straw and the root of the plant will be ploughed in to keep up the humus level for subsequent crops.

Lime (in the form of ground chalk) being spread by contractor.

Sustainable Sources of Energy

Modern farming, just like modern living, uses far more energy than in olden days in the form of electricity, diesel fuel for tractors, and oil for the production of fertilizer and crop protection chemicals. The world's scientists and politicians will have to develop much better sources of energy than the wind generators, which are heavily subsidised and only work in windy weather, to ensure that there is a sustainable future source of energy.

Sustainable World Population

The population of the world continues to increase, so more mouths will have to be fed in the future, and it will be the duty of farmers to grow enough food for all those extra mouths. In order to maintain sustainable world food supplies farms will have to be farmed intensively. As a local example on our farms, in 1955 we were pleased to grow just over one tonne of wheat per acre, but now we are disappointed if we grow less than three tonnes per acre. There will also have to be less food wastage in processing, catering and private homes, and we will have to use the latest techniques of genetic modification (GM) to increase crop yields further still.

Forestry Commission Woodland. Chilterns Commons. Purchase of Ditchfield Common by Lane End Parish Council.

Forestry Commission contractor harvesting softwood, 60 years after planting.

Forestry Commission Woodland

The report in February this year that the Minister for Defra (the Department for the Environment, Food and Rural Affairs) has decided against selling off the nation's woodland owned by the Forestry Commission was good news to most of us who live in the rural areas. There had been widespread criticism at the suggestion that the State could no longer afford to own or manage its own forests. The Forestry Commission had been formed in 1919 after the demands of trench warfare in World War I had used up so much timber that there were no remaining reserves of wood either in store or growing in Britain's forests. The aim of the new Forestry Commission was to promote forestry and the production of timber. It was the timber thus grown that became invaluable for the needs of World War II, at which time the Commission clear felled 29,530 acres of its forests to use the timber at a time when imports were not possible.

In the 1950's and 1960's most of this forestry land was replanted with softwoods, fast growing coniferous species such as Larch, Pines, Norway Spruce, Douglas Fir or Western Hemlock from which the timber is used in building construction for roof trusses, doors and window frames. Many folk did not like the view of these new plantations, some of which were planted in rectangular blocks which were out of harmony with the natural contours and valleys of the hills on which they were planted. This area of forest land owned by the Forestry Commission was enormous, nearly 4 million acres by the start of the 1970's.

However, from the 1970's up to the present time the Commission has changed its policy objective away from commercial production of softwood timber on a big scale in all of its forests towards conservation and amenity in some areas like our own Chiltern Hills, where most of the woods are privately owned with the Commission only owning about one tenth of this woodland. The Commission now tries to maintain woodland character and recognizes the importance of broadleaved trees, such as oak, ash or cherry which grow slowly but naturally to produce hard timber with attractive grain suitable for making furniture.

The beech trees which grow naturally in our local Chiltern Hills woodland, on which the High Wycombe furniture making trade was based in earlier years, come into this category of broadleaved deciduous trees, shedding their leaves every Autumn and growing slowly to produce quality trees which are likely to take up to 120 years to mature, before they are ready to fell for the furniture making industry.

Chilterns Commons

In and around Lane End we are lucky to have eight different Commons which, just like all Forestry Commission woodland, can be enjoyed and walked over by the general public. These are described in more detail in the guidebook which the late Peter Philp prepared in the Millennium Year 2000, entitled "Four Pubs and Two Duck Ponds - Lane End, The Birth and Growth of a Chiltern Village".

Many people do not realize that Common land is not owned by the State as public land, although some commons (like Wimbledon Common or Burnham Beeches) are owned by public authorities or by boards of conservators or trustees for the benefit of the general public. Other commons such as Ashridge Common in Hertfordshire are owned by the National Trust, but the freehold of common land in rural areas is normally owned privately, and often the grazing rights are still important to livestock farms.

In Lane End the Parish Council already owns most of Moor Common, Moorend Common and the small Oakshaw Common off Simmons Way. These Commons, and the five commons in private ownership, are all registered in the Commons Register held by the Buckinghamshire County Council and all of the owners have permitted the general public to walk over their commons at all times within living memory.

During the 19th Century the commons in Lane End would have been used for local industry rather than recreation, or as open downland grazed by cattle and sheep, at a time before the motor car had been invented. Cadmore End Common was owned by Lord Parmoor until the 1940's, and right up until 1938 was used for quarrying the materials from which the local bricks were made. Clay was dug from open pits on the common, chalk was quarried from deep shafts below adjoining parts of the common, and the brick molds were filled by hand

and wheeled on small rail tracks right into the kiln, which still stands to this day. Fuel for baking the bricks was straw from adjoining farms and firewood grown on the common.

Purchase of Ditchfield Common by Lane End Parish Council

Ditchfield Common is in private ownership at the present time, but the Executors of the late owner, Tom Taylor, have offered it for sale in recent weeks to Lane End Parish Council. The Parochial Church Council of Holy Trinity Church, which stands in the middle of Ditchfield Common, feels that Ditichfield Common is so important for the village of Lane End that the freehold of this common land should be owned and controlled as a community asset, by Lane End Parish Council.

The Church Council have therefore set up a special fund through which Parishioners can make donations to the Church for this purpose. The intention is for the Church Council to lend the sizeable sum in this fund to the LE Parish Council, to enable the Parish Council to buy Ditchfield Common straightaway. Then the LE Parish Council would repay this loan to the Church Council over a ten-year period on interest free terms.

The Parish Council will also be pleased to receive direct donations, either from individuals or from associations, towards its proposed purchase of the freehold of Ditchfield Common so that it will be able to own and control Ditchfield Common as a community asset for the benefit of all the people of Lane End at all future times.

Residents fear their common could be bought by travellers

A MUCH-loved open common in Lane End has been put up for sale following the death of a London landowner – prompting fears it could eventually become a travellers' site.

A meeting on Tuesday heard Ditchfield Common has been put up for sale by the widow of the late Thomas Taylor – a solicitor from Highgate who had inherited the land. Though owned by the Taylor estate, Ditchfield is common land, meaning others hold traditional rights of use or access, such as to pasture 50 cattle and 25 sheep on the land.

The common has been used for recreation by villagers for many years and is viewed by many as a public asset. Holy Trinity Church sits on its own plot within the common, and church warden Bryan Edgley, **pictured**, has urged villagers to club together and help the parish council buy the land. He told about 50 residents at the Youth and Community Centre on Tuesday: "The parish council I'm absolutely certain ought to own it to have the future management of the common.

"Most of us remember the last invasion of travellers. If Ditchfield Common is owned by an unscrupulous owner they could sell it to gypsies. The encampment would be there in the middle of the village."

The ten-acre common is listed at £75,000, but Mr Edgley believes an offer of about £50,000 may be accepted.

Parish council chairman Graeme Coulter said the site has 'no economic value' due to its common land restrictions, but warned someone could buy it in the hope that Commons laws may change. He said it was unlikely the council would buy the common without significant financial support from residents, as it would involve raising the parish tax.

A show of hands at the end of the meeting showed unanimous support for a council-led purchase of the land, with almost everyone saying they would be willing to contribute financially.

One woman said the issue was of greater importance to the village than the proposals for a community stadium at Wycombe Air Park. She added: "If the heart of the village goes we may as well have the stadium as well – and I'm absolutely anti the stadium."

Mr Edgley told the **Free Press** he was delighted with the passionate response from villagers. He stressed the Taylor family have been 'immensely helpful' to the village and church, saying they had gifted part of the common for a churchyard at Holy Trinity in 1997.

Villagers were asked to email the parish clerk if they are interested in making a financial contribution. Email: clerk@laneendparishcouncil.org.uk

Report in the Bucks Free Press of 29th April 2011, when Ditchfield Common was on the market, that "Holy Trinity Churchwarden Bryan Edgley has urged villagers to club together and help the parish council to buy the land".

Note: During the months following publication of this article the Lane End Parish Council purchased the freehold of Ditchfield Common using finance from its own resources, without any loan from the Parochial Church Council of Holy Trinity Church. The money raised by parishioners of Holy Trinity Church will thus be available for maintenance or improvement of Holy Trinity Church, in which the author serves as one of the two Chruchwardens.

The Clarion – *Autumn 2011*

Heatwave in Early Summer Months. Market Value of Grain. Grain Sales by Openfield, a Farmer Owned Cooperative. Chinese Lanterns.

Wheat harvest - the Golden Ball of St. Lawrence Church, West Wycombe can be seen on West Wycombe Hill in the background.

Heatwave in Early Summer Months

I was asked this morning, on the last day of July, how the harvest work is progressing, and what effect the early heatwave and lack of rain in the Spring has had on the yield on our wheat crops? The answer has to be that in this area the excessive rain in June was a real benefit to the cereal growing crops, even though it caused immense difficulty to livestock farmers attempting to make silage or hay at that time.

In the three months of March, April and May total rainfall was only 91 millimeters so we were expecting the crops to ripen too early with very poor yields. But then rain in the month of June was 98 millimeters, similar rainfall to a wet winter month, which came just in time for the wheat to resume growth. We started harvest on 28th July, only a few days earlier than usual, and yields so far have averaged at about 3 tonnes per acre. This is slightly less than most years, but still satisfactory, whereas in the eastern counties such as Kent the June rainfall came too late to improve those crops which had suffered from the early season drought.

Market Value of Grain

The farm's gross sales from the arable fields depend as much on market value of grain as on the yield per acre. And market value is not controlled locally, since wheat is an international commodity. It can be shipped at fairly low cost from one country to another, and so wheat yields in America or Russia will be just as important as yields in South Bucks when the world's grain traders balance the likely world stocks against anticipated worldwide demand. And worldwide demand will include wheat that is distilled to make bioethanol as a source of energy, as well as wheat used by the human race for bread, biscuits and breakfast cereals and also wheat used to make livestock feeds for pigs, poultry and cattle.

There are then further difficulties with this international commodity trade in that some poor countries have insufficient funds to buy enough grain to to avoid starvation amongst its people, and at the other end of the scale some grain trading companies, such as Cargill and Associated British Foods in developed countries have become so large that they can now manipulate the market to their shareholders' advantage.

Grain Sales by Openfield, a Farmer Owned Cooperative

On the farm we have to accept the going rate for the grain, most of which we sell through a farmers' co-operative called Openfield. For the year 2009/10 worldwide consumption of wheat was 262 million tonnes, but estimated production was only 244 million tonnes, so worldwide reserves were depleted, leaving a buffer of only 180 million tonnes. Thus the market reacted to this shortage between demand and supply with an increase in the worldwide market value. This was good news for those of us who grow wheat, but bad news for farmers who have to buy feeding stuffs for their livestock. For the housewife there should be no further increase in the price of bread, since the ingredient cost of farm grown wheat is only a small proportion of the retail cost of a loaf of bread, generally between 12% and 25% dependent on brand, and shop prices have already increased in recent months to take account of this higher market value of wheat.

Chinese Lanterns

Will Lacey writes on a different page of this issue about the real menace and danger to the health of cattle from 'Chinese lanterns' which have landed on pasture fields. Farmers hope that these Chinese lanterns will be made illegal, since they are a fire risk to ripe cereal crops and to buildings as well as causing the death of cattle. Our Member of Parliament, Steve Baker, visited Laceys' Family Farm in June where we were able to discuss with him the need for better contracts and prices for milk producers and the need to control badgers infected with tuberculosis as well as our hope that Chinese lanterns will be made illegal.

The Clarion – *Winter 2011*

Michaelmas Day. Solar Photovoltaic Panels. Blackgrass and Sterile Brome Grass.

Michaelmas Day

The Autumn quarter starts with Michaelmas Day, the 29th September, which is traditionally the end of the farming year, a time when the current year's cereal crops of wheat barley or oats were safely in the stackyard and planting for next harvest, to be ready in August of the following year, had not yet started. It was also a logical end of year date for the livestock farms, when Summer grazing on the meadows was nearly over and Winter feeding, of livestock with the fodder which had been produced in the growing season, was about to start.

The modern equivalent of Michaelmas Day is one day later, on the last day of September, and that is still the financial year end for most farms. So it is also a good time to review the past season and to plan for any changes or new projects for the next year.

Solar Photovoltaic Panels

At Kensham Farm one project that we hope to bring into use before the end of March 2012 is the installation of Solar PV panels on two of our barn roofs. This is in line with current Government initiatives for producing energy from natural resources. These photovoltaic panels are a more effective way of producing electricity than wind generators - which often spoil the landscape and do not produce any electricity unless the wind is blowing at the right speed.

We have already been granted planning consent to have these Solar PV panels fixed on aluminum rails above the south facing roof slopes of one of our grain stores which is fifteen years old and a fertilizer store which is two years old. It would be no good to fix them above an old roof in need of repair, nor to fix them on a north facing roof slope. For technically minded readers of this column, it is hoped that the 138 poly-crystalline Amerisolar panels, each with dimensions of 1650mm x 991mm, will generate 235 watts each at peak output to give a total installation output in bright sunshine of 32.43 kilowatts.

It will be a long term project, with a planned life of not less than twenty five years, during which time we will have a two way connection to the National Grid. This will be wired up so that electricity generated by the daylight and sun shining on the Solar PV panels on our barn roofs which is surplus to our own requirements will be exported back into the National Grid on a feed-in tariff at an inflation linked and Government guaranteed price. We hope that it will all turn out as planned, and that the panels will still be in good working order in twenty five years time - but that remains to be seen!

Solar panels on a barn roof being power washed after 5 years use.

Blackgrass and Sterile Brome Grass

Our Autumn seeding, of crops which will be harvested in August 2012, was completed in excellent working conditions by the third week in October. The dry weather in September and October was really good for working the soil down to friable seedbeds, enabling the seed drill to plant the seeds into a good tilth at an even depth. We always take care not to allow invasive grass weeds, such as Black Grass or Sterile Brome Grass, to gain a foothold in the wheat crops. Our usual methods of control are firstly to work down the stubbles immediately after harvest, so that any weed seeds are encouraged to germinate. The young grass weed plants can then be killed with a total weed killer such as Roundup, which works on the leaf of the growing weed but becomes inactivated and safe when it contacts the soil. The second control method is to spray the field with a 'Pre-emergence' spray, sprayed onto the soil immediately after the crop has been planted but before it has germinated. The young weed plants will then be killed as they emerge through the treated soil, but the good wheat plants will emerge safely since they are resistant to the effect of the pre-emergence spray. The only snag in a dry year is that pre-emergence sprays do need moisture in the soil to work effectively - so, as with most of farming, it is a balance between extremes of weather that is best.

The Clarion - *Spring 2012*

Farm Work in Winter. World Grain Trade. Imports of GM Soyabean. Genetic Engineering. Chilterns Woodland Project.

Farm Work in Winter

The late winter months on an arable farm are the time for maintenance work, such as machinery overhauls, cutting back overhanging trees and trimming hedges while we wait for the drier conditions and warmer soil temperatures, generally in March, when the first applications of fertilizer can be spread on the fields.

The main productive work over the winter has been checking the condition of grain in grainstores and loading lorries with wheat. This is the milling wheat to be used for making bread and biscuits, that was harvested in August 2011, from fields that had been seeded in September or October 2010, with seeds that had been ordered in May 2010 following a cropping plan that we had worked out in Autumn 2009. So farming is not a trade where next week's production will be governed by this week's sales.

World Grain Trade

Farmers have to accept the ups and downs of the world grain trade. So if there have been bumper yields of grain in countries as far away as Australia, Russia or America then world supply will be in excess of demand, as it was in January 2010 when feed wheat traded at £95 per tonne. However, if there has been drought and crop failures in other parts of the world, causing a shortage of grain on the world market, then the same quality of feed wheat could trade at more than double the price, such as the £200 per tonne that it was worth in April 2011, but by December 2011 this had fallen back to £135 per tonne.

On the subject of global food security, that is whether or not there will be enough food for a growing world population, many scientists think that genetically modified foods (GM foods) will have to be used more widely in the future. In the USA 93% of soyabeans are now grown as GM crops, whereas in 1996 when this technology was being developed only 5% of the USA soya was genetically engineered. Soyabeans are a very important vegetable protein, used in most poultry, pig and cattle feeds throughout the world, but the crop can only be grown in hot climates.

Most of the soya imported into the UK for animal feeds has been genetically modified, since very little traditionally bred soya has been grown in recent years. But in Europe there has been a fierce anti-GM food campaign, led in the UK by environmental groups, egged on by some newspapers. But these campaigners, who talk of "Frankenstein foods", do not take note of the life saving advantages of some GM foods and plants in third world countries, such as golden rice in which the Vitamin A content is boosted, or varieties of crops which are resistant to drought, or to excessive salt, or insect repelling crops which do not then have to be sprayed with insecticide. One example of this has been genetically modified cotton in India which has increased yields by 60% by making the crop unattractive to bollworm.

Genetic Engineering

Crop plants have been improved by selective breeding for many thousands of years, and progress has been faster in the period since the end of World War II in which crop yields have trebled, so that where one tonne per acre was grown in the 1940's we now grow over three tonnes. Genetic engineering is different to selective breeding, in that by using the soil bug Agrobacterium desirable traits from an unrelated species of plant can be introduced into the species of plant that needs improvement e.g. for greater drought resistance, or improved nutritional benefits, or a tolerance to specific weed killers, or more resistance to pests or diseases.

So the big question is whether genetic engineering of plants is good or bad. Many of us think that this advance in science, developed during the past fifteen or twenty years by global agro-chemical manufacturers, will be necessary to ensure that sufficient food is grown for the increasing numbers of humans in the world. But we wish that it was being developed by University departments, or by Government backed research farms, rather than by multi-national commercial firms who seek to profit from these developments.

Felling trees on Bolter End Common - a job for which traffic lights had to be used to control road traffic.

Chilterns Woodland Project

Getting back to Lane End, and laying aside the world's problems, readers of the Clarion V may notice trees being felled in Finings Wood and Bolter End Common in the coming months. This is no cause for concern, since it follows the woodland improvement scheme prepared by John Morris, Director of the Chilterns Woodland Project for which we have already received Forestry Commission approval and Felling Licence. It will involve a woodland contractor felling about 420 of the trees which are dead, spindly, unhealthy or overcrowded and a few which are too close to Finings Road. We hope that this will benefit all the best trees during the years to come.

The Clarion - *Summer 2012*

Irrigation and Abstraction Licences. Saddle Gall Midge Larvae. Oilseed Rape.

Hosepipe Ban

Rainfall this season has been brought to everyone's attention by the current hosepipe ban in March, followed by an exceptionally wet April, after a winter when the reservoirs and ground water reserves were not sufficiently replenished by natural rainfall. Gardeners can still water their plants and vegetable gardens with a watering can instead of a hosepipe, but all the local farms on these Chiltern Hills have to rely on natural rainfall, since mains water would be too costly to be economic for use on cereal crops or on pastures for cattle or sheep.

Irrigation and Abstraction Licenses

Those farmers in other regions who use irrigation have to use a natural source of water, such as a river or borehole from which water can be pumped, and generally then they only use irrigation on high value crops such as potatoes or vegetable crops. But even if they own the bed of the river from which they want to pump water for irrigation they are still only allowed to pump out the water if they have a Water Abstraction Licence granted by the Environment Agency. In south east England demand for water is so great that new Abstraction Licences are seldom granted, and in the current season further restrictions have been imposed on long standing existing Abstraction Licences.

Saddle Gall Midge Larvae

At Kensham Farm we have kept accurate rainfall figures throughout the last twenty years. These records show that April 2012 has been the wettest April in those twenty years, with 152 millimeters (just over 6 inches) recorded in the month. This followed three unusually dry winter months, January to March, in which total rainfall was only 112 millimeters. This high April rainfall has been good rather than bad, but it has reduced the number of days suitable for spreading fertilizers or spraying crop protection chemicals. However, it has been a great benefit in limiting the hatch of Saddle Gall Midge larvae, which only hatch when soil conditions are dry and warm (over 18 degrees centigrade). These larvae feed on the stem of the growing wheat plant, thereby weakening it and reducing the yield of grain. In the last two years of hot dry springs we experienced a significant reduction in the yield of wheat from some of our fields, despite three spray treatments of insecticide, but this year the heavy and continuous rain in April seems to have provided a most effective control.

Oil seed rape in flower.

Oilseed Rape

This season we are growing oilseed rape on some of those fields on which the midges had caused trouble last year, having taken that decision in late summer last year, long before we realized that it would be such a wet April this season. Oilseed rape fields can always be recognized when the fields become bright yellow in April and May as the rape plants come into flower. Seed pods are then formed containing the very small round seeds, about two millimeters in diameter, which are the valuable part of the crop.

The oilseed rape crop is harvested generally early in August with the same combine harvester which harvests wheat and other cereal crops. However, the trailers carting the rape seeds to the grain dryer have to be well maintained, almost watertight, otherwise the tiny seeds would pour out from any small crack or badly fitting tail gate. Sale of the rape seed is through a grain merchant, in our case through the large Farmers' Co-operative trading under the name of Openfield from offices near Andover in Hampshire, to one of the only two firms which have crushing and refining facilities in the UK.

At the refinery the rape seeds are crushed to extract the oil, and the remaining fibrous part of the seed is made into meal as a by-product suitable for feeding to livestock. The valuable rapeseed oil (known in the rest of the world as canola oil) will mostly be made into margarine, mayonnaise, ice cream or cooking oil. But some of the rapeseed oil will be used as a lubricant, and some will be used as a biofuel which can be mixed with mineral oil for use in diesel engines.

162

The Clarion - *Autumn 2012*

Milk Supply Chain. Farm Subsidies. Set-Aside Land Brought back into Cropping.

Milk Supply Chain

The most important farm topic in recent months has been the failure of the milk supply chain, resulting in dairy farmers losing money at a time when supermarkets are drawing a healthy profit from retailing the milk. About 20% of output from British farms is dairy produce, so it is important for the whole of the farming industry for this sector of farming to be making a profit.

Will Lacey will say more about this on his page, but readers of Clarion V will remember that last year our MP, Steve Baker, visited Bolter End Farm and discussed with us the better contracts and milk prices that dairy farmers will need to cover their costs and to make a profit with which to pay for their household living expenses.

At Kensham Farm we sold our dairy herd over forty years ago and turned our cattle yards into grain stores. But it is good for us to remember that a dairy farmer has to own or rent a farm, on which he has to grow grass and fodder crops for the cows to eat, and he has to buy or breed the cows and look after them, and call in the vet for any routine tests or treatment, and he has to house the cows and feed them in a yard or stalls throughout the winter. He then has to build a milking parlour and milk the cows in it at least twice each day for seven days a week including Christmas Day and the other bank holidays. It seems quite wrong that the dairy farmer should lose money for all this hard work and investment, whereas the major retail chains, which between them control 88% of all food sales in England, have sufficient marketing power to maintain their customary margin of profit on the milk which has been produced at a loss by the dairy farmer. We look forward to seeing whether the proposed Adjudicator, who will be responsible for upholding the Groceries Supply Code of Practice (GSCOP), will be given sufficient powers to correct this failure of the milk marketing chain.

Farm Subsidies

I am sometimes asked how farm subsidies work, and whether farmers still have to set aside some of their land. Farm subsidies originate from the EU, and in recent years the total of these payments received by British farms has been very similar to the total profit from farming. In other words, on average farm produce is sold at the farm gate at cost of production, and it is only the EU payments that give the farmer a profit enabling him to continue trading. So the real beneficiary of the EU farm payments is the consumer, who is able to buy all the food for the household at the bare cost of production.

These EU payments were significantly changed in 1993 and again in 2005, in that in earlier years the payments were linked to production, whereas for the last seven years the EU Single Farm Payment has been paid on an acreage basis (measured in hectares) irrespective of the amount of food being produced, and is really a State payment for farmers to look after the land. Under this scheme land can still be set aside, but that is no longer compulsory.

Further State payments are also made to farmers for wild life conservation. At Kensham Farms we have signed a second five-year contract under the Entry Level Scheme (ELS) for carrying out these environmental works such as planting wild bird cover strips, leaving field margins and corners, keeping paddocks in natural permanent grass, and cutting the hedges in alternate years to give birds more cover and food than if the hedges were to be cut every year.

Sumo heavy cultivator breaking up permanent pasture for its return to arable cropping.

Set-Aside Land Brought back into Cropping

We have recently carried out some contract work on land that had been in set aside for over twenty years. Bringing this land back into production, to grow a crop of wheat ready for harvest 2013, has been an interesting task. The first job was to kill out the grass and weeds on the set aside land with glyphosate ('Roundup') followed by deep cultivation, then a dressing of 20 tonnes to the acre of organic manure, which is a processed material from the sewage works, immediately buried by further cultivation. We then seeded a cover crop of mustard, which will both comply with the EU regulations to seed a crop within six weeks of spreading the organic manure, and will also benefit the soil by suppressing weeds and providing additional organic material, and by giving off a gas from its roots which will diminish potential wireworm damage to the next cereal crop. This mustard cover crop will be sprayed off and then incorporated into the soil in September, to make the field ready for seeding with winter wheat to be harvested eleven months later in August 2013.

The Clarion - *Winter 2012*

Spread of Bovine Tuberculosis. Compulsory Slaughter of Cattle infected with Bovine Tuberculosis. Trial Cull of Infected Badgers.

Spread of Bovine Tuberculosis

Rural areas are under threat from bovine tuberculosis (bTB) spread by diseased badgers, and in my opinion this threat to the countryside is more serious than any other problem in the rural areas at the present time.

We are not affected by the spread of bTB by infected badgers at Kensham Farm, since we do not keep any dairy cows. But this threat to British livestock farming is so serious that I want all of our readers of Clarion V to understand why the cull of badgers, proposed for two specific pilot areas of West Somerset and Gloucestershire, continues to be so important even though it has had to be postponed from October 2012 until Summer 2013.

The series of six maps shown in the picture below was produced by Defra in 2011. It shows how a few isolated cases of bovine TB found in cattle during routine tuberculosis tests carried out by Ministry Veterinary Surgeons in 1986 had become endemic in most of the western parts of England and South Wales by 2010.

Compulsory Slaughter of Cattle infected with Bovine Tuberculosis

The captions below the map shows that 28,451 cattle were compulsorily slaughtered on Government orders in 2010 and since then, in the latest full year of 2011 the number of cattle tested positive for bovine TB had increased to 34,000. All of these cattle are now dead, since as soon as a cow fails the tuberculosis test by the Ministry Vets it has to be shot on the farm, and its carcass has to be incinerated. One West Country farmer who witnessed one of his favourite dairy cows being shot following Ministry orders, after failing the test for bTB, could see the baby calf in its womb kicking and struggling for life inside its mother's dead body before it suffocated. And even if it had been born alive before its mother had been killed, the baby calf would still have had to be shot, since its mother had failed the bTB test.

There is an upset to the balance of nature, whereby the present excessive number of badgers is now decimating the population of ground nesting birds such as lapwings, grey partridges, larks and willow warbles. The badgers are also killing bumble bees and hedgehogs – leaving just their prickles uneaten.

Badgers were seldom seen thirty years ago, although there were well established and healthy badger setts, like inter-connecting tunnels in the earth, in woodland areas. Then Government introduced the Badgers Acts of 1973 and 1991, which were then consolidated into the Protection of Badgers Act 1992. This has resulted in such an increased badger population that 50,000 badgers were killed in road accidents in 2011. They now have to compete with each other for food to the extent of digging up lawns in private gardens when looking for earthworms to eat, and some colonies of badgers have dug new setts in which to live in open fields and pastures.

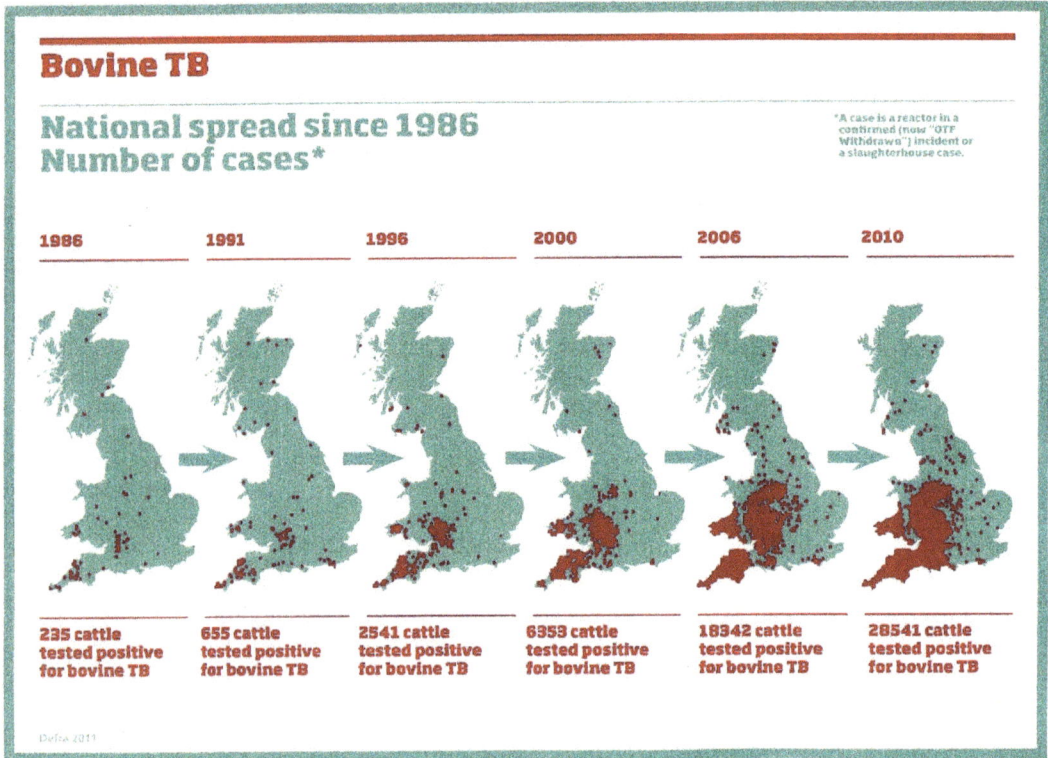

Bovine TB

National spread since 1986
Number of cases*

*A case is a reactor in a confirmed (now "OTF Withdrawn") incident or a slaughterhouse case.

| 1986 | 1991 | 1996 | 2000 | 2006 | 2010 |

| 235 cattle tested positive for bovine TB | 655 cattle tested positive for bovine TB | 2541 cattle tested positive for bovine TB | 6353 cattle tested positive for bovine TB | 18342 cattle tested positive for bovine TB | 28541 cattle tested positive for bovine TB |

Defra 2011

The spread of bovine tuberculosis from 1986 to 2010.

Trial Cull of Infected Badgers

Tuberculosis in the human race was a fatal disease before World War II, until the discovery and development of antibiotics which could cure the disease. It is therefore essential for Government to continue to test cattle to ensure that neither the cattle nor the milk produced by the dairy cows can carry any TB infection. In this respect the new Minister for Defra, the Rt Hon. Owen Patterson MP whose family farm in Shropshire, is determined to bring in the proposal trial cull of badgers as the first step towards a good control policy. He is also working on a new polymerase chain reaction test (PCR) which will be able to ascertain from faeces found near a badger sett whether or not the badgers in that sett are infected with bTB. Research scientists at the University of Warwick have shown this test to have an accuracy of 97-98%, so that it could lead to control measures specifically targeted at infected badgers.

There is no suitable vaccine to protect cows from the disease, since current vaccines for cattle also trigger a positive reaction to the tuberculosis test. Research chemists are working on the new 'Deva' vaccine, which does not trigger a positive reaction to the tuberculosis test but so far it is only 46% to 55% effective at protecting the cattle.

Some journalists suggest protecting the badgers with a vaccine, but this would not cure any of the badgers carrying the disease (in 2009/2010 29% of road killed badgers were found at post mortem examination to be infected with bTB) and anyway it would be wholly impractical task to attempt to treat wild animals such as badgers in the same way as farmed animals like cattle or domesticated animals like dogs.

Many folk, and many urban politicians, think that badgers are cute and cuddly – but a closer reading of Beatrix Potter's novel '*The Tale of Mr. Toad*' reveals the truth of the badger's character – in which Tommy Brock is described as being "not nice in his habits" and steals a sack full of baby rabbits to eat.

Farmers want to see a return to a healthy balance of wildlife within the countryside, with sensible numbers of healthy badgers as well as all the other species of wild animals. They do not want to live under the threat of their life's work, and the care of the livestock in their charge being destroyed by bTB contracted from diseased badgers.

The Clarion – *Spring 2013*

Rainfall Records. Difficulties in a Wet Autumn. Felling Roadside Trees. Mobile Sawmill.

Rainfall Records

Spring is a good time to review the past year, and let us hope for much less rain in 2013. In 2012 the rainfall records that we keep at Kensham Farm showed an increase of 34% above an average year, with a total of 1,104 millimetres (43.5 inches) whereas our ten-year average, 2002 to 2011, showed an average of 821 mm (32.3 inches).

Harvest 2012 was a difficult time, with all the grain having to go over the grain dryer, some of it three times, in order to bring it down to between 14% and 15% moisture content. If we had not dried the grain it would have not been saleable and would not have kept well in store. Damp grain in store spontaneously heats up, just like fresh lawn mowings on a compost heap, and then mould sets in.

During August, while we were still harvesting, we had successfully seeded the fields to be cropped with oil seed rape in August 2012. Since then we have used whirling bird scarers, which spin round in the wind, as well as some white plastic bags on tall sticks, to keep the pigeons away from the young plants.

But then the wet Autumn caused even more difficulties for the next arable cropping season, since the fields were too wet to take the weight of a tractor for the Autumn seeding. In a normal year the first seedings of wheat, barley and oats are drilled in the middle of September, and we hope to complete seeding by the end of October. These cereal crops would then be ready for harvesting in August of the following year.

However, the fields were so wet throughout September and October that we were only able to seed three quarters of the planned acreage, and some of that has subsequently failed due to slugs eating off the young cereal plants at ground level. If Summer had been normal, with plenty of sunshine, then the grain used for seed would have been plumper and that might have given it the vigour to withstand some slug damage.

We were fortunate over seeding the fields near Harecramp, off Chequers Lane, with winter wheat on 16th January – after two fairly dry weeks and with enough frost to be able to take the tractors onto the land we were able to cultivate, harrow and seed the fields, finishing at 2.0 am the next morning. Mid-January is the latest that winter wheat can be seeded, so those fields that were too wet to seed in the Autumn and the crops destroyed by slugs will all have to be seeded with Spring variety crops in February, March or April.

Cultivating at Harecramp, off Chequers Lane, in January following two dry weeks. Often the fields are too wet from November until early March to take the weight of a tractor.

Felling Roadside Trees

One winter job that went well was the work on Finings Wood, each side of the B482 Finings Road on the Stokenchurch side of Lane End. This work involved felling those trees which, following advice of the Chilterns Woodland Project, were considered to be dead, spindly, unhealthy or overcrowded and some which were too close to the road. This work was carried out by woodland contractors using a 'cherry picker', a hydraulic working platform with a reach of 27 metres mounted on a lorry but having all the control leavers in the raised working platform. The traffic on the B482 made the job more difficult, necessitating the use of traffic lights.

Some longstanding Lane End residents have told me how they have appreciated seeing the result of this work, especially on the downhill side nearest to Finings Farm, and that the restored wood is now as they remembered it as children. We have sold most of the cordwood arising from this felling to be collected either by local residents with trailers or by firewood contractors. The ash logs burn well when green, whereas logs from oak and other species such as beech, birch or hawthorn burn best after drying out in a stack for a season.

Cherry picker with 27 metres reach cutting down dangerous trees, close to the the B482 Finings Road at Lane End.

Mobile Sawmill

Two of the best oak trunks were sold to a local woodcarver for use as signs, carved nameplates and other craft work. These trunks were cut into suitable widths and cross sections in lengths of about 10 foot each with a mobile sawmill – a machine which comes in sections in a truck or trailer and is then assembled on the site. Finings wood was so wet in early December that the truck had to be parked by the roadside and all the sections had to be carried up to the spot where the oak trees had fallen. It was fortunate that this work coincided with a study day on Woodland Management on Commons run by the Chilterns Commons Network, so that the owners and managers of other Commons were able to see this contractor's mobile sawmill at work, converting locally grown oak into timber for local craft use.

Mobile sawmill planking oak tree trunks. With this system the mobile sawmill can be transported with a Land Rover and trailer to the wood, when the tree trunk is maneuvered with crowbars onto the sawmill.

The Clarion – *Summer 2013*

Beamish Living Museum of the North. Seed Drill Mechanism. Poor Establishment of Wheat in a Wet Autumn.

Historic tram on its tramlines at the Beaulieu Living Museum.

Arable tramlines in winter wheat crops at Kensham Farm.

Beamish Living Museum of the North

On a recent visit to the Beamish Living Museum of the North, between Durham and Newcastle upon Tyne, we used the tramway which has been built at the Museum to visit the different exhibits on the 300-acre site. One of the exhibits was the 'Home Farm', a small farmstead as it would have been in late Victorian or Edwardian times. This was a reminder to me of my time learning the farming trade at Sacrewell in 1953, when my job during the sugar beet harvest was to harness up one of the seven working horses, then work with it all day carting the beet in a cart that could carry just one ton.

How farming has changed, where the present-day equivalents are a tractor, most often developing between 100 and 360 horsepower, from a diesel engine of between 4 litres and 9 litres cubic capacity, pulling a trailer with load capacity of twelve to sixteen tonnes, or a plough with five to ten furrows.

The historic tramway rails at the Beamish Museum reminded me of the origin of the 'tramlines' which are used nowadays by the tractors for all present-day cereal crop treatments, so that the tractor with sprayer or fertiliser spreader can carry out the work without crushing the crop.

Seed Drill Mechanism

Most crops of cereals, mainly wheat but also barley and oats, are planted as winter crops in September and October. At seeding time we use a seed drill with hopper taking three tonnes of seed, sufficient to drill about 50 acres from each fill of the hopper, covering a total width of 8 metres, with 52 disc coulters at 15cm centres. The tractor drawing the drill is equipped with electronic controls receiving GPS satellite signals, connected to a computer with a screen monitor in the cab showing the driver the size and overall shape of the field being drilled. This computer is linked directly to the steering on the tractor, which can then steer itself exceedingly accurately, so as not to miss out any land nor overlap between one bout and the next, but the tractor driver can override the computer by taking the steering wheel. The monitor shows the position of the tractor and seed drill in the field, the line it is following, the area which has been drilled in one colour and the area not yet drilled in a different colour.

The tractor computer has a sophisticated connection to the seeding mechanism on the drill, to blank off the seed supply to each pair of rows intended to be used as tramlines. The computer cuts off the seed to three coulters for each tramline row, this matching the width of a tractor rear wheel. We always use tramlines at 24 metre (80 feet) centres, so that the first time round the field all the rows of drill coulters for the full drill width of 8 metres will be sowing seeds, then the second time around the field the coulters corresponding to the width of the tramlines will be automatically blanked off, thus leaving two unsown rows at exactly the correct width for the tractor which will carry out subsequent crop treatments. Then on the third time around the field all the coulters will be sowing seeds. In this way the unseeded tramlines left across the field will all be precisely 24 metres apart, the correct working width for the crop sprayer and the spreader which will be used after crop emergence and for all subsequent crop protection treatments during the growing season.

Poor Establishment of Wheat in a Wet Autumn

All readers of The Clarion will have already experienced the excessively wet Autumn 2012, then the long cold winter resulting in Spring 2013 being around four weeks later than a normal year. This is the reason why the crops in early Summer this year are looking so disappointing, many with bare patches on which the Autumn planted seeds failed to germinate properly or were eaten off by slugs as they emerged. Some of our fields of Autumn sown wheat were so poor and uneven that in March or April we sprayed off whatever few plants remained, then cultivated the field to form a new seedbed and re-sowed with a Spring variety of wheat, or on some fields replaced the wheat with Spring barley or oats or oil seed rape. We must now hope for favourable weather in the months before harvest, with plenty of sunshine interspersed by sufficient rainfall for crop growth.

This year our new grain dryer and store is being constructed on the small field between our existing farm buildings and the M40 Motorway, and we hope that it will be ready before harvest.

Open Farm Sunday will be on 9th June when we hope to be at Bolter End Farm supporting the Lacey Family who are hosting it this year.

GPS monitor in John Deere tractor showing position of tractor with seed drill in the field, and area already seeded.

Harvesting Methods. Arable Crops

The modern method of harvesting. The John Deere combine harvester with 9 metre, 24ft, cutter bar finishing harvest on Sunters field, West Wycombe Estate.

Harvest

August is the main harvest month for cereal cops, but some years, and further North in England, it is not completed until September. Modern harvesting machinery relies on big machines, run by a relatively small staff.

The harvest photo was taken last harvest, August 2012, of the John Deere S680i combine harvester powered by a 13.5 litre diesel engine developing 550 brake horsepower which we now use at Kensham Farms. This carries out the same task as the self-binder with two horses and the threshing machine run by the steam traction engine with a team of nine men, in that it cuts the stems of wheat and passes them with the bushing auger on the cutter head into the threshing drum inside the combine harvester to separate the grain from the straw. The combine harvester steers itself on all the straight runs by means of satellite navigation, with the route, the area already cut and position of the combine in the field, the yield, the moisture content and any troubles such as grain not being separated correctly all being shown on the computer monitor screen in the cab on the combine, constantly monitored by the combine driver.

Arable Crops – Wheat, Barley, Oats and Oilseed Rape

The crops for this year's harvest were mainly planted in August to October as Autumn seedings with one block of fields being seeded in January as a result of the wet Autumn. Then the Spring seedings, using different varieties of seed corn, were seeded between late February and early April. Most years we grow winter wheat for milling on most of the fields, but this year, following the wet ground conditions in 2012, we have also grown spring wheat, spring barley (used for malting to brew beer, or for animal feeds) oats (used for porridge, muesli or horse food) and oilseed rape, which is then crushed to extract the oil to be used for making margarine, mayonnaise, ice cream, cooking oil and some as a lubricant or fuel for diesel engines.

The combine harvester has a grain tank that holds about 7 tonnes of grain with a sensor that flashes an amber signal lamp on the cab roof whenever it is nearly full. A tractor drawing a grain trailer with a capacity of up to 16 tonnes is then driven alongside the combine which discharges the grain into the trailer while the combine is still at work cutting the corn, with the two machines travelling at exactly the same speed and the tractor driver taking care to keep the trailer level with the combine and the correct distance away to receive all the grain coming out of the combine discharge auger.

Whenever grain is moved nowadays it is moved in bulk using hydraulic or electric power for lifting. A hundred years ago the average small mixed farm would have been pleased to grow forty or fifty tonnes of grain in a whole year but the combine harvester which we use normally harvests between forty and fifty tonnes each hour and can actually achieve one tonne of grain harvested each minute when all the conditions are just right.

The Clarion - *Winter 2013*

CAP Reform. MacSharry Reforms. The Single Farm Payment. Pillar 1 and Pillar 2 Payment.

CAP Reform – what does CAP stand for and why is it being reformed?

At the present time the European Union (EU) in Brussels is carrying out a periodic re-appraisal of the Common Agricultural Policy (CAP) with a view to proposed changes coming into force in 2014 – changes which will have a real effect on the viability and living standards of every farmer in Europe, including Britain.

The Common Market was established when the Treaty of Rome was signed in 1957 by its six founder Member Nations. One of its early objectives was to solve the food shortages, with UK rationing of food, that had started in World War II and which continued for another ten years after the end of the war. Discussions in the European Community (the EC as it was named at that time) were led by France and Germany and resulted in the formulation of the Common Agricultural Policy (CAP) in 1962.

The European Union controlled British farming from 1972 until the Brexit vote in 2020.

The CAP was then modified in 1968 under the Mansholt Plan, with a new social and structural plan for European agriculture. All of these early schemes to end food shortages, and to prevent any recurring shortage in future years, were so successful that by the 1980s there was overproduction of food with surpluses in Europe, leading to grain and butter 'mountains' and 'wine lakes' which had to be stored in EU Intervention Stores. This was the time when milk quotas and 'Set Aside' were introduced to curtail food production in the Europe.

MacSharry Reforms 1992

The MacSharry Reforms in 1992 were then brought in with further alterations to the CAP designed to move its provisions towards a free market. Under these reforms agricultural subsidies were linked more to looking after the natural environment, instead of being solely linked to food production. In this context Governments were aware that farmers also maintain the rural landscape as a by-product of productive farming.

A tidy and attractive countryside benefits rural folk, but it also benefits the tourist industry with overseas visitors, and the many people living in towns who increasingly want to spend their leisure time in the countryside. Thus Government can have all the 'park keeping' of the farmed areas looked after by farmers, at no extra cost to the Government, by ensuring that the farm businesses are viable and thriving.

The Single Farm Payment

The next big change to the CAP was in 2003, when the 'Single Farm Payment' was introduced by the EU to sever the final link between subsidy and the food produced from the land. This EU subsidy, administered in England by the Rural Payments Agency (RPA), is a flat rate payment to farmers for maintaining the land in cultivable condition. Under the scheme farmers also have to maintain good standards, called 'Cross Compliance', for maintaining the natural environment with its wildlife, food safety and animal welfare – with many of the requirements imposed by UK Government being higher than in some of the other EU Member States.

The current proposals for re-appraisal and periodic reform of the CAP hope to reduce the cost of support and to address the difficulties in recent years of instability in farm prices, which now vary wildly from season to season dependant on whether there is a perceived world surplus or world shortage of raw food ingredients such as wheat for making bread. The majority EU view is that the CAP at EU level is still more desirable than a series of national or regional policies for agriculture, but recognises that there are variations in conditions between the 27 different countries which are now Member States of the EU.

Pillar 1 and Pillar 2 Payments

The CAP regulations describe payments made direct to farmers as 'Pillar 1 Payments'. These are straightforward to administer (providing the farmer and the RPA both attend to the SFP Claim Form efficiently!), whereas rural development grants, described as 'Pillar 2 Payments', involve such an immense amount of red tape that successful claimants are likely to be those organisations or charities which are good at filling in application forms, rather than local management and initiatives which could provide good outcomes for the countryside, the soil, water resources, and the wildlife within them without an excessive administrative burden. The movement of available funding from 'Pillar 1' to 'Pillar 2' is known, in EU parlance, as 'Modulation'.

Farmers hope that the Department for the Environment, Food and Rural Affairs (the unfortunate name under which the old Ministry for Agriculture now operates), when it considers the responses to its present consultation, will remember the importance of enough food being grown on the farms in this country rather than relying on imported food. This means that sufficient funds should still be allocated to 'Pillar 1' to keep farmers in business, rather than so much of the diminished EU funding being allocated to other non-farming environmental measures through 'Pillar 2' as to put farm businesses at risk. We farmers have had to remind the Minister for Defra that such a change would put English farmers at a disadvantage compared to farmers in other EU Member States, and that consumers do not want the shop price of food to go up.

Last year's excessively wet autumn made several of our fields too wet to take the weight of a tractor for the autumn seeding. But this autumn conditions for seeding our wheat and barley have been excellent and were completed before the end of October. These young cereal crops are now around four inches high, looking healthy and so off to a good start for harvest 2014. We sprayed some of them in November with insecticide as a precaution against aphids with our new Amazone self-propelled sprayer, shown in the photo at work on Chequers Manor Top Plain, next to the B482 road to Stokenchurch.

Amazone self propelled sprayer.

The Clarion – *Spring 2014*

RASE Medal for Nigel Rogers. Plant Protection Chemicals. Specialised Farm Production. Flooding on the Somerset Levels.

RASE Medal for Kensham Farms Foreman

What is special about the date 18th February 1962? All readers of Clarion V who take an interest in modern history and have a knack of relating dates to how old they were, or how many years ago it was, will straightaway recognise this date as having been fifty-two years ago. But for us at Kensham Farms it just happens to be the date when our farms foreman, Nigel Rogers, started working here.

So at the time of writing these jottings I have nearly completed preparation of a PowerPoint presentation showing changes such as the change from milk and egg production on our farms to arable crop production, and the increased size and complexity of the tractors and farm machinery, over those years. We hope to show this to Nigel and his family, including his son Paul who is taking on Nigel's duties as foreman, at a supper to celebrate his 52 years of work for Kensham Farms - and to hand over a commemorative medal presented to him by the Royal Agricultural Society of England.

Nigel Rogers, Foreman at Kensham Farms since 1962 with his son Paul, during harvest 2013.

Plant Protection Chemicals

Thinking of the many changes in the farming industry which have taken place since 1962, one of the most significant has been the advance of science leading to plant protection chemicals. These include fungicides to protect the young growing cereal plants from diseases such as mildew or septoria, growth regulators which reduce the length of straw and increase both the strength of the stems and the root structure of growing cereal plants. This avoids the difficulties which we used to get from lodged crops - crops which had laid down flat on the ground before harvest because they had outgrown their own strength.

Research scientists have also developed effective herbicides, such as the weed killers which are selective so that broad leaved weeds, like docks, stinging nettles, charlock, poppies, and cleavers ("sticky willy"), can all be killed with a crop spray which causes no damage to the growing cereal plant. A further development of these weed killers has been in the formulation of "pre-emergence" sprays - herbicides which are sprayed on the bare earth immediately after seeding, so that seedlings of weeds will be killed as they poke up out of the ground through this invisible crust of treated earth - and yet the cereal plants can grow up through this crust without harm.

At a later stage of growth cereal plants can now be sprayed with an insecticide to control pests such as aphids or saddle gall midge, which could otherwise cause immense damage to the developing plants. However these insecticides will not be necessary in every season, and will only be used if routine inspections have revealed a real threat to the growing crop.

Winter wheat is seeded in September or October. The young plants establish before the cold winter months, then remain at about the height shown in this photograph until growth starts again in February or early March.

180

Specialised Farm Production

Further major changes, prompted by these improved crop protection chemicals, by farm economics and by the change to a global food market, have led to more specialised farm production. So nowadays there are very few mixed farms of a hundred or two hundred acres, the size of the average farm which existed in this part of the Chiltern Hills fifty years ago. Farm size of the majority of serious food producing farms has greatly increased, while other smaller farms with insufficient production to provide a living wage for a farmer have become part-time farms. These small holdings are owned and run by those who have farming and livestock husbandry in their blood, and are able to supplement their farm income with diversified interests such as a farm shop, providing bed and breakfast or holiday accommodation, or even having additional income from sources away from the farm.

Flooding on the Somerset Levels

Turning now to the excessive rainfall over the past winter months, our sympathy must go to farmers and others in flat and low-lying rural areas such as the Somerset levels, where flooding has been compounded by Government and Environment Agency shortcomings in failing to carry out sufficient routine dredging and maintenance of the main waterways such as the River Parrett in recent years. Almost unbelievably, the Environment Agency now regard the habitat and wellbeing of a water vole as being more important than keeping the rivers flowing properly.

We are fortunate in this area in living on hills overlying chalk, with such good drainage that most Chiltern valleys are dry, without ditches or streams, since the rainfall percolates through the chalk into the underground aquifers from which our water supplies are pumped up.

Our crops this season were mainly seeded into excellent dry and friable seed beds in the dry months of September and October, and the seed was plump and with good vigour following Summer 2013 with plenty of sunshine, so that our crops are now well established with strong young green plants which will not have been harmed by the heavy rain in December and January.

The Clarion - *Summer 2014*

Pastures damaged by Flooding. Benefit of a Dry Harvest and Autumn. Open Farm Sunday. New Grain Dryer and Grain Store

Pasture damaged by Flooding

A few months ago flooding was the main concern of folk who lived anywhere near a river and its flood plain. You may remember that the Spring issue of Clarion V, only three months ago, carried the headline "Marlow under Water" and showed an aerial photo of Higginson Park under water.

At that time some villages and farms in the Somerset Levels were marooned for weeks on end, leaving farmers in that area in a terrible situation from which they have not yet recovered. All the productive species of grass in those river meadows such as Perennial Ryegrass and Timothy were killed off whereas unproductive pasture plants such as couch grass with its underground rhizomes, and buttercups, have regrown from the reserves in their root structure once the flood water drained off. Furthermore, the earth worms which are beneficial to soil structure have also been killed off by the floodwater. So these farms are left with most unproductive pasture, which should really be killed off with a total weed killer such as glyphosate, sprayed on the leaf but which kills the root of the plant as well as the leaf, and then re-seeded.

However, on our Chiltern Hills soil, which is mainly clay with flints over lying chalk, it has been an excellent season so far for growing wheat. The critically important months of August and September 2013 were dry, so that harvest was straight forward and only a modest amount of grain drying was necessary. Following cultivation the dry and friable seedbeds were just right, so that the new seeds for the wheat crop, which will be harvested in August 2014, germinated well and became strong healthy plants before the winter really started.

When the rainfall was excessive from December to February it was important not to take a tractor anywhere near the fields for crop treatments, and it was only a very few areas of low-lying land where the crop suffered from poor drainage. We then had favourable weather with low rainfall in March and April for carrying out the crop treatments of fertilizer spreading, and spraying with herbicide weed killer, fungicides to protect against plant diseases such as mildew and septoria, and growth regulator to make the straw stronger but shorter and to strengthen the root system.

Open Farm Sunday

At Kensham Farm on Open Farm Sunday, 8th June, we will be open to the public from 2:30pm until 5:00pm. We hope that readers of Clarion V with their families will come to see our crops growing, and have a ride on the tractor drawn trailers, and see the farm machinery with which the land is cultivated and prepared, and the seed drill with which the crops are seeded, and the fertilizer spreader and self-propelled sprayer used for applying crop protection sprays - and then climb up into the cab of the combine harvester to sit on the driver's seat and see the controls.

Open Farm Sunday at Kensham Farms. The tours around the farms are always popular.

Tea and refreshments will be provided by the Friends of Cadmore End School, for which donations will be appreciated towards a new computer server that is needed by the school. The event itself will be free of charge, but donations will be invited for LEAF, the 'Linking Environment and Farming' association which co-ordinates Open Farm Sunday and shows details on its website (www.farmsunday.org) of the 400 farms which will be open to the public on Sunday 8th June.

Laceys Family Farm at Bolter End are involved in organising the Young Farmers County Show this year, so will not have any displays for Open Farm Sunday - but instead they will provide a display of produce, and some of their Guernsey calves, here at Kensham Farm.

New Grain Dryer and Grain Store

Those readers who travel on the M40 Motorway will probably have seen our new grain store which was built in time for the 2013 harvest. Visitors on Open Farm Sunday will be able to see this grain dryer, which can dry 46 tonnes per hour of wet grain, and the grain store which holds 3,000 tonnes of wheat. The photo shows the grain store and the tree guards of the 200 trees which we planted in March as a shelter belt. These trees are a mixture of sycamore, wild cherry and small leaved lime together with a few of other species like field maple and whitebeam and including evergreen species such as holly and Scots pine. The young trees on the motorway bund are only 40 to 60cms in height, but these should take well and establish to provide a good screen in the years to come.

The M40 motorway was constructed across the land at Kensham Farm and opened for use in 1967. The Kensham Farm grain dryer and grain store can be seen on the right.

The Clarion - *Autumn 2014*

Michaelmas Day. Early Harvest. Variable World Price of Wheat. Incorporation of Straw into Soil.

The John Deere combine harvester at dusk. Combining will often continue until 11.00pm, sometimes even later if there is no evening dew.

Michaelmas Day

September is a good time to review the past farming year and Michaelmas Day, the 29th September, was the day when traditionally any change in tenancy or ownership of a farm took place. The logical reason for this was that the cereal harvest and the main summer grazing season for livestock had ended, and the winter seeding for the following harvest, or the yarding of cattle for winter feeding, had not yet started. Most farms still use Michaelmas as the end of the farm financial year for accounts purposes, although to suit modern computer systems it is generally delayed by one day to the 30th September.

This year the cereal harvest started earlier than in most years. Our combine harvester started cutting the winter barley crop on 15th July and completed it on 17th July. Barley straw has a useful feed value for cattle, as well as making good bedding for cattle yards, so for those fields we disconnected the straw chopping mechanism at the back of the combine harvester, to leave the straw in swathes. We then sold it in the swathe for use by the cattle at Laceys Family Farm at Bolter End, for which it was baled and removed from the field straight away. The importance of baling and carting the straw without delay for us is so that Autumn cultivation can start straight away in preparation for the next season's harvest, and for the livestock farmer is so that the straw can be gathered in before any rain which would spoil its quality.

The combine harvester can be set to chop the straw, or to let it pass through if it is to be baled as in the photograph.

Variable World Price of Wheat

Our wheat harvest started on 27th July, and the excellent yields this year will partly counteract the fall in value per tonne of wheat on the worldwide market. It always seems surprising that a staple food such as wheat for making bread should have such volatile changes in value. Tables published in Farmers Weekly in August show a comparison of two year's prices for feeding quality wheat to have varied from a high point of £214 per tonne in November 2013 to the present low point of £112 per tonne in July 2014, this variation being the result of the laws of supply and demand. If the season for crop growth has been favourable both in this country and in other wheat producing countries, as it has been for harvest 2014, then the world's grain markets will judge that so much wheat is in store globally as to amount to a worldwide over supply. When that happens farmers still have to sell the grain that they have grown, and so will take a lower price to clear it even if that lower price is less than cost of production. However, another favourable factor this year will be a saving on the cost of fuel oil and electricity for grain drying, since much of the grain in the early part of harvest came in from the field sufficiently dry, at 13% to 14% moisture content, so that no further drying was necessary.

Incorporation of Straw into the Soil

The straw from most of our wheat crops has been returned to the field to be incorporated in the soil as organic matter for the next crop. This is achieved by rotating blades which chop the straw into lengths of two or three inches before it leaves the rear end of the combine harvester, and then a fan spreads it evenly over the thirty feet wide cutting width of the combine header.

Our combine harvester has been achieving an output of around fifty tonnes per hour, and on the best day we were able to harvest 600 tonnes. This has meant no time to waste in carting the grain back from the harvest field to the grain stores in the trailers carrying 14 to 16 tonnes per load. Some readers of Clarion V may remember seeing one of our tractors with a flat tyre and a trailer behind it parked in the village outside the Golden Palace and McColl's paper shop and store in August. Thanks to those who were patient over any disruption caused by this breakdown while the tyre was repaired. But that was not quite the end of the story, since tractor tyres are made to flex to grip better on fieldwork – but internally fitted repair patches do not flex quite so much. So unfortunately the tyre repair did not last, we had to replace all four tyres on that tractor and the quotations for the cost of the four tyres varied from £13,000 plus VAT for the most expensive brand down to £5,000 plus VAT for supplying the four Indian BKT Agrimax tyres which are shown in the photo being fitted onto the John Deere tractor back in our farm workshop.

Fitting 4 new tyres to the John Deere is a task for two fitters.

The Clarion - *Winter 2014*

Berkshire College of Agriculture.

Berkshire College of Agriculture (BCA)

Many farmers from South Buckinghamshire and Berkshire attended an Open Day at BCA in November, the title of the presentation that we heard was: "Putting the A back into BCA". The Berkshire College of Agriculture (BCA) at Burchetts Green near Maidenhead was founded just after World War II, at a time when food rationing was still in place - and everyone in Britain, whether in the towns or in the rural areas, had no doubt that a reliable supply of food was essential.

However, in recent years school leavers have often been discouraged by their careers teachers from going into agriculture, regarding farming as being an old-fashioned activity, no longer necessary in the modern world of progress, sophisticated IT systems and globalisation. How wrong they are. Food production is as essential as it ever was, but the labour requirement on the farms is no longer for a large staff of strong unskilled workers - instead the work of growing food is now carried out by a small number of farmers with highly skilled workers. Arable farms now use powerful tractors with GPS monitor screens in the cab, and the livestock farms often use computer technology to record weights and yields and control nutritional needs of the large flocks and herds in their charge.

At BCA misunderstandings about modern farming and lack of interest by schoolteachers caused the demand for places at agricultural colleges to diminish, to the extent that many well-established colleges closed. BCA closed down its farming courses in 2001 and for the past thirteen years have only taught allied subjects such as horticulture, ground care, wildlife conservation and equitation as land based industries, together with other subjects such as caring for children.

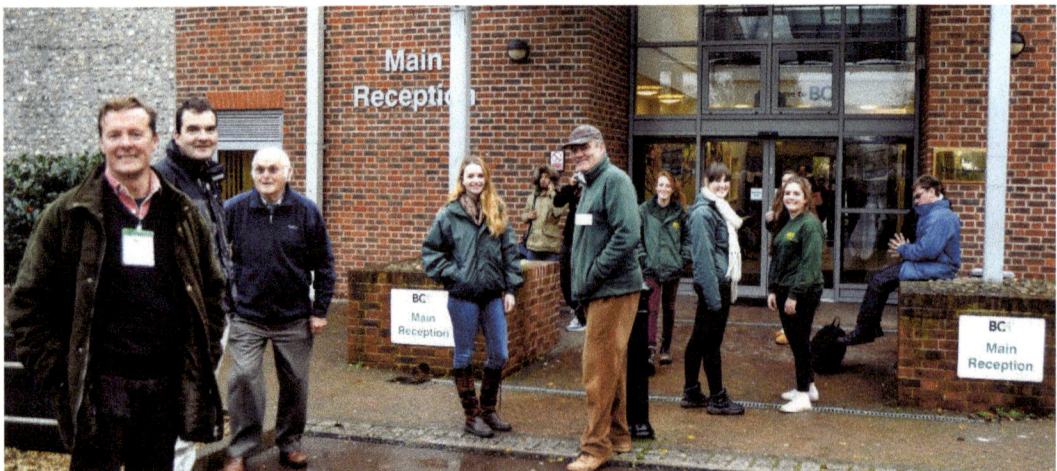

Delegation of NFU members visiting BCA. Charlie is on the left.

In recent years youngsters wishing to follow their time at school with a college course on farming have had no options in this area, but now BCA has resumed its basic farming tuition with a Level 3 Extended Diploma in Agriculture, taught as a two-year course with three full days of teaching each week combined with work experience. In the first year the youngsters learn about livestock and crop production, anatomy, physiology, plants and soil science. The second year of the course is centred more on the business aspects of farming, going into the detail of dairy, beef and sheep production as well as forage crops, organic farming, wildlife habitat and pollution and waste control management.

This year there is an intake of 18 students on the farming course, seven lads and eleven girls, some of whom showed us round the college grounds - their interest in learning the skills needed in the modern farming industry and their enthusiasm was a delight to some of us older farmers who attended this BCA reintroduction of its farming courses.

Charlie talking to students studying Agriculture at BCA, the Berkshire College of Agriculture.

Spring Drilling. Vernalisation of Winter Wheat. Importance of Good Seedbed. Malting Barley.

Spring Drilling

Spring Drilling is the first important activity on arable farms following the winter, when the fields are sufficiently dry to take the weight of a tractor with the seed drill. The photo, of seeding spring barley in West Wycombe Park, was taken in the second week of March 2014.

Much of the farmland around Lane End is used for growing wheat, barley, oats or oil seed rape, arable crops which are harvested with a combine harvester in August and early September. Wheat is the main crop, used for milling into flour to make bread, biscuits and some breakfast cereal such as Shredded Wheat or Weetabix and All Bran (which is made from wheat, using just the husk from the outside of the wheat grains). In the flour milling process the white kernel inside the grain is separated from the coarser skin of the wheat grains to make flour for white bread, whereas with 'wholemeal' bread the whole of the wheat grain is used, without separating the husk from the inner kernel.

Spring seeding of wheat in West Wycombe Park in March. The residential area of Downley, High Wycombe is in the background.

Vernalisation of Winter Wheat

Wheat is always seeded either in the Autumn, from the middle of September to the end of October, or in the Spring, normally in February or March but in a wet Spring this may be delayed until early April. The winter wheat varieties are divided into the milling wheat varieties with high protein content and feed wheat varieties with lower protein, but higher yields, for feeding livestock. The winter wheat varieties need a period of 'vernalisation', that is cold weather on the growing plant before it forms seed heads. If winter wheat seeds were to be planted in the Spring they might not form any seed heads until the following season.

Spring wheat has a shorter growing season with no requirement for vernalisation, but normally does not provide such high yields as winter wheat. So winter wheat may have a growing season of eleven months, whereas spring wheat may only be in the ground for five months between seeding and harvest, but they will both be ready to harvest at a similar time in August.

Importance of Good Seedbed

Soil condition at seeding time is just as important on the farm as in a garden. The ideal seedbed will have been formed by cultivations so that it is fine, with no clods or cavities, and firm enough to take the weight of the seed drill. The seed drill shown in the photos has a working width of 8 metres, has 52 individual seed coulters at spacings of 15 centimetres through which the seed is planted to a depth of about 4 centimetres. The seed drill carries up to 3 tonnes of seeds, which will have been dressed with a coating of seed dressing to guard against such soil borne diseases as seedling blight and eyespot.

Accurate drilling has been simplified in recent years by GPS technology, so that the tractor can steer itself parallel to the previous bout, and precisely 8 metres from it, with a monitor screen in the tractor cab showing the shape of the field, the position of the drill in the field with the area already drilled in a different colour to the remainder of the field.

Malting Barley

Barley and oats have similar growing requirements but are generally less profitable than wheat to grow. The best quality barley will be used for malting - the main ingredient for brewing beer, which is flavoured with hops. Surprisingly, this malting barley will be the lowest in protein - whereas the best quality wheat for milling will have the highest protein, as well as a good score for gluten. Gluten in necessary to make the loaf of bread rise when baked, and can be measured in a laboratory test to calibrate the 'Hagberg Falling Number' of the sample in question.

Best quality oats will be used for the production of muesli, porridge and oat cakes, whereas the poorer samples of oats, as well as lower grade wheat and barley, will be used for the production of feeding stuffs for pigs, poultry and cattle.

Oilseed rape is often used as a break crop in an arable rotation of crops. It is seeded either in August or during April or May, comes into its bright yellow colour when flowering in the Summer, and is generally ready for harvesting with the combine harvester in July. The rape seeds will be crushed to release oil, to be used for the manufacture of cooking oil, mayonnaise, margarine and ice cream.

Growth Regulator for Cereal Crops. Importance of Politics for the Farming Industry. DNA Testing of Badgers. Food Production or Conservation?

Growth Regulator for Cereal Crops

It is now the end of one of the driest April months on record, with our cereal crops looking good, but we are hoping for some rain soon to keep them growing. We have already sprayed the wheat crop twice with a tank mixture containing primarily a fungicide to protect the crop from mildew and diseases such as septoria which affects the leaf and eyespot which affects the stem. The other ingredient of the tank mix is a growth regulator to improve the root strength and shorten the eventual height of the straw - so that it will remain standing upright, that is without "lodging", at harvest time - even if it turns out to be a wet summer. During May and early June we will treat the crops twice more with fungicide, hoping to keep the leaf a healthy green colour right up until it starts to ripen in July.

Amozone Pantera self-propelled sprayer. The boom width is 24 metres (80ft), here seen spraying wheat with fungicide and growth regulator.

Importance of Politics for the Farming Industry

In ten days there will be an important General Election - the result will be known to everyone by the time this Summer issue of Clarion V is printed. Politics are enormously important to the farming industry.

Badgers are protected at law under the Protection of Badgers Act 1992, so that it is illegal for a farmer to put a diseased badger out of its misery without Ministry consent. But all livestock farmers, especially in the West Country, will hope that the new Government will improve methods of controlling bovine tuberculosis by controlling the disease in wildlife. The connection between infected wild badgers and incidence of the disease in cattle has been proved by DNA testing of mycobacteria taken from the lungs of road killed badgers, which have been found to correspond to the same sub-type of mycobacteria as cattle which have failed the tuberculosis test in the same Parish.

DNA Testing of hairs and faeces of badgers

Many of us agree with a remark made by Princess Anne on Countryfile, that the best method of control would be humane gassing of badger setts which were known to be infected with bovine tuberculosis. The good news is that scientists are already developing DNA testing of hairs and faeces of badgers that can be found close to a badger sett. When this DNA testing is fully proved it will be possible to ascertain whether or not the sett is infected, so that the only setts to be gassed could be those which are known to be the home of badgers already ill and dying from bTB.

The severity of bovine tuberculosis is such that in the 12 months of the 2014 year 32,851 cattle were compulsorily slaughtered in the UK after failing the tuberculosis test, this being even worse than in 2013 when 32,612 cattle failed the test - and each test failure was a real personal worry, as well as financial loss, to the farmers and stockmen who had been looking after the cattle. The disease has spread towards the East from the West Country so that in the Chiltern Hills, we are now considered to be an area at risk, with all cattle in Buckinghamshire having to be tested by Ministry Vets for tuberculosis every year, rather than once every four years.

Food Production or Conservation?

Another political outlook which is of huge importance is whether the farms should concentrate on the production of food for human consumption, or whether conservation of wildlife should continue to be given more emphasis in the UK than food production. The Economist magazine published an article earlier this year entitled "Dig for Victory" in which it stated that many British farmers 'just trundle along', without the author recognising that it was politicians, not farmers, who had changed their main objective for farms in England from efficient food production to conservation.

Farmers in England have responded to this changed political outlook with massive changes in the structure of farms in England over the past two or three decades, resulting in the contraction of generalist farms, to become small hobby or lifestyle farms for those with supplementary sources of income, and the expansion of those fewer remaining serious food producing farms with additional land - occasionally by purchase but more often by renting land from neighbouring holdings, or by "share farming", or by farming additional land on a contract basis. So we farmers have to keep a watchful eye on the politicians as well as on the crops and livestock on our farms.

The Clarion – *Autumn 2015*

Wheat Harvest. Computer Monitoring on the Combine Harvester. International Trading of Wheat as a Commodity. Diamond Wedding.

Traffic on Local Roads

Congestion of traffic on local roads is likely to have been caused in recent weeks more by tractors than by parents taking their youngsters to school, for this is the busiest time of year for arable farms - with harvesting work being immediately followed by Autumn seeding.

John Deere combine harvester harvesting wheat. The straw is being chopped by the combine to be incorporated back into the soil to maintain the humus content for the next season.

Wheat Harvest

The photograph was taken during the first week of August, when we were harvesting wheat on our field known as Niddles, just off Bigmore Lane and bounded along one edge by the M40 motorway. This field, with the adjoining field next to Leygroves Wood, has shown up the different soil types in Summer 2015 with its unusually low rainfall – whereby the wheat on the poor parts of the fields ripened earlier and with lower yields than on the rest of the field. The total rainfall shown by our rain gauge at Kensham Farm has averaged 865mm (34 inches) per year over the past ten years, but over the four months from March to June 2015, in which the crops put on most of their growth, rainfall was only 168mm (6.6 inches).

The combine harvester is fitted with a computer operated monitoring system, which continuously checks that all the thrashing drum clearances and sieve settings are set to minimise grain losses, and it also continuously measures the yield of grain. This has accurately measured the different yield between that on the best parts of the fields, in which the soil type is predominately clay with humus able to retain moisture, and the other parts of the field with lighter soils, largely sand or gravel, which dried out too early this summer. On the best parts of Niddles, shown in the photo, the yield was up to 4 tonnes per acre of milling wheat and on the Motorway Field, on the right of the road from Lane End to Sands, yields reached nearly 5 tonnes per acre, whereas on the more gravely parts of the Bigmore Farm fields the yield was only 2 tonnes per acre, following premature ripening due to lack of retained moisture in the light soil.

Profit from the crops will be influenced by growing conditions, but even more by the state of the global market place. Wheat can be exported to other countries quite cheaply by sea, and so large tonnages are traded internationally. The value of the wheat which we sell is controlled by the worldwide supply and demand situation, rather than by English weather affecting home grown crops. Wheat sold by us through our farmer owned and controlled grain marketing co-operative, Openfield Ltd, had peaked at about £220 per tonne in 2012/13 but is currently only worth £120 per tonne.

International Trading of Wheat as a Commodity

These peaks and troughs in the global market are shown on the graph produced by the Agriculture and Horticulture Development Board and the Home-Grown Cereals Authority of UK farm gate prices prepared by LIFFE - the London International Financial Futures and Options Exchange, which is the electronic trading platform formed in 1998 for international trading of commodities such as grain. This high finance of international commodity trading is a far cry from our efforts in actually growing wheat on these fields in the Chiltern Hills, but it is effectively the factor which determines whether we will make a profit, or incur a loss, from growing crops on our farm.

Similarly, the value of milk sold on the wholesale market in the UK is now below the cost of production, a situation which can only carry on for a limited time. Current political thought is that free market conditions will balance supply with demand – but many of us farmers have doubts that this will provide long term stability of food supplies which have been produced to the high standards of British farm produce. If consumers feel the same, then please make a point of buying British locally produced food whenever possible.

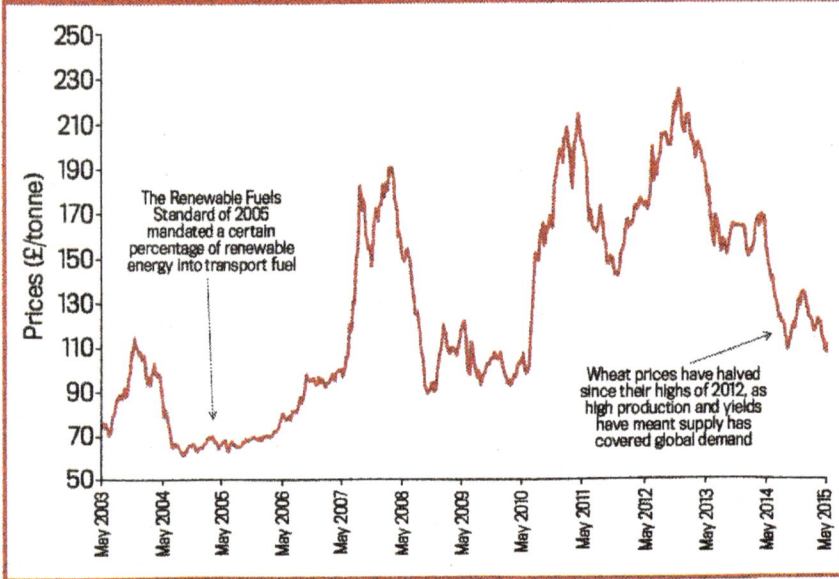

Nearby LIFFE wheat prices — Source: AHDB/HGCA

The Renewable Fuels Standard of 2005 mandated a certain percentage of renewable energy into transport fuel

Wheat prices have halved since their highs of 2012, as high production and yields have meant supply has covered global demand

Chart published by the Agriculture and Horticulture Development Brand of statistics compiled by the London International Financial Futures and Options Exchange (LIFFE) showing the fluctuations in the farmgate price of wheat over the 12 year period 2002 to 2015.

Diamond Wedding

Turning now to more personal matters, this issue of Clarion V is likely to land on readers' doormats in September 2015, which will be just 60 years since Alison and I married and started farming at Kensham Farm. And remember that in 1955 if you took a £1 note to the garage when you filled the car up with petrol you would have been supplied with 4 gallons (18 litres) of petrol and still have received some loose change.

Bryan and Alison at their Diamond Wedding with Charlie (left), Paul and Polly.

Crop Rotation. Development of Hormone Herbicides. Use of Glyphosate.

Crop Rotation

Some readers of Clarion V who have glanced toward the Motorway when driving from Lane End to Sands will have noticed the large field on the right, formerly part of Grove Farm, on which a newly planted crop appeared to emerge in September in stripes across the field, and then in October part of the field nearest to Sands was ploughed up and drilled again with fresh seeds. So what happened?

Before World War II the standard method of controlling weeds and diseases in arable crops was to use a rotation of crops. The "Norfolk Four-Course Rotation" was the most widely used rotation – it had been developed following the Enclosures of land in the 1800s and was in general use until the introduction of artificial manure, and the Agriculture Holdings Act 1948.

This rotation had been Roots (turnips, swedes or mangolds) in year 1, followed by Barley in year 2, then Seeds (largely red clover) for year 3 followed by Wheat in year 4. In this way any cereal diseases would have been controlled in the alternate years by the Roots or the Seeds which were not susceptible to such diseases, and weeds such as thistles or charlock would have chitted and been destroyed in the Root crop, either with a horse-drawn hoe, or with hoeing in the same way that a gardener would use a long-handled hoe in the spring or early Summer, between the rows of the root crops. The folding of sheep within temporary hurdles on the Seeds and Roots, and the winter feeding of bullocks in yards, were essential parts of that system of farming.

Young oil seed rape plants.

197

Development of Hormone Herbicides

Then artificial fertilisers became more widely used, and during the Second World War research by scientists both in the UK and the USA led to the development of the first modern herbicide, named 2,4-D, a growth stimulant which would kill certain broad-leaved weeds by making them outgrow their own strength without harming the cereal crop in which they were growing. These weed killers were far more effective at controlling broad leaved weeds than the control achieved by hoeing the root crop of the Norfolk Four-Course Rotation. So it became possible to grow consecutive crops of cereals, yields increased when competition from weeds no longer used up the available plant nutrients, and it was no longer necessary to keep to a rigid system of crop rotation to control weeds and plant diseases.

At this time in the 1940s and the 1950s production of home-grown food was essential for the nation since there had been a shortage of food in Britain during World War II, which was controlled by Ration Books continuing into the 1950s. Research scientists continued their work in developing plant protection chemicals with improvements in seed dressings and fungicides – for controlling such diseases as blight, septoria and eyespot in wheat and controlling mildew on the leaves of other crops, both on the farm and in the garden on susceptible plants such as roses as well as vegetables. Effective insecticides were also developed which could kill pests such as the Cabbage Stem Flea Beetle, and also aphids without killing the ladybirds which feed on the aphids.

Use of Glyphosate

In 1974 Glyphosate ('Roundup') was introduced for non-selective weed control to kill the leaf of any plant, and translocate down its roots, and yet be neutralised as soon as it touches the earth. Then in the 1970s Graminicides were developed, these are herbicides designed to kill grass weeds without harm to broad leaved crops such as oilseed rape in which the grass weeds may be growing. Graminicides also kill 'volunteer' cereals - that is young plants of wheat, barley or oats which are not wanted, having grown from grains which were not harvested properly by the combine harvester.

On most arable fields we like to start cultivations or ploughing as soon as the crop has been harvested. In this way any volunteers or weeds will have time to grow and then be sprayed off with a total weedkiller such as Roundup. Then the field can be seeded with the new crop seeds into a weed free seedbed.

So with this brief history of plant protection treatments the troubles on our Motorway Field in Autumn 2015 can be explained. Oilseed rape has to be seeded in August, which leaves no time for chitting and killing off any volunteers or weeds in this way – instead we rely on spraying the young oilseed rape crop with a graminicide for control.

So the green stripes on Motorway field were volunteer plants of wheat, which have been subsequently killed off with the graminicide. And the section which has been ploughed nearest to the wood, shown being re-planted with oats in the photo, was the result of bird damage – young oil seed rape plants, just as they emerge from the ground, make a tasty meal for birds.

Section of field being seeded with oats in October, following bird damage to oil seed rape in the remainder of Motorway Field.

The Clarion – *Spring 2016*

Winter Work on the Farm. Future of UK farming if we leave the EU. Farm Gate prices for Farm Produce. Care of the British Countryside. Badgers and Bovine Tuberculosis.

Winter Work on the Farm

What do farmers do in the Winter? At Kensham Farm we do not have livestock jobs such as feeding and bedding down cattle, milking dairy cows or the shepherd's task of assisting the ewes to give birth to the baby lambs at lambing time, which is generally February to early April each year.

Instead of that one of our main productive tasks in the Winter is looking after the grain in store and loading lorries. In late January, when there was a ship in Bristol Docks being loaded with milling wheat destined for Algeria, we loaded 48 articulated lorries in one week, each carrying just under thirty tonnes. On one day Charlie loaded 6 of these lorries, that is about 180 tonnes of wheat, before breakfast. Loading is carried out with the Manitou rough terrain telescopic forklift, shown in the picture with its bucket, which holds 2 tonnes of wheat.

Loading a 30 tonne capacity lorry with wheat at the Kensham Farm grain store.

Winter is always a good time of year to take a strategic look at the farm as a business, before the Spring and start of the growing season. At the end of January our Wycombe Constituency MP, Steve Baker, spent the whole of a Friday afternoon with us and nine other local farmers and NFU advisers discussing the main farming concerns both now and in the near future. Some of these issues, with Steve Baker's views shown as 'SB', are: -

The future for UK farming if we were to leave the EU

SB is strongly in favour of Brexit, that is the UK withdrawing from membership of the EU altogether. If this were to be approved at the proposed Referendum, he would not expect any immediate changes, but would expect a relaxation of regulations to take place gradually. SB realises the benefit of relatively cheap food in the shops which is possible by subsidising the production costs of farmers, and would expect a similar system to continue – perhaps better administered and targeted than at present through the EU. On the lighter side of this hugely important subject, and in the context of recent policies formulated by the EU, SB remarked that even a broken watch does show the right time twice each day.

Steve Baker MP with Alison and Bryan following his off-road ride around the Kensham Farm tracks on his BMW motorcycle - prior to more serious discussion of the future of UK farming.

Farm Gate prices for Farm Produce in 2015/2016

SB realises that as a result of two better than average growing seasons throughout the whole world, there is now an international surplus of wheat and milk. This surplus, following the laws of supply and demand, has resulted in these food commodities trading at less than cost of production. Recent statistics published by HSBC confirm that position. But no political control has yet been devised to solve this volatility.

Care of the British Countryside

Farmers are as keen as any other rural folk to look after the countryside and the wildlife that lives in it, and a majority of farmers had joined the EU 'Entry Level Scheme' for specific measures such as Wildflower Strips to benefit birds and insect life in field edges during the past ten years. But the EU and UK Government have now taken the inept decision to scrap that scheme, and to replace it with a badly planned new scheme which will spoil much of the benefit to the countryside that followed from the last ten years of ELS care for the countryside.

Badgers and Bovine Tuberculosis

We were pleased to hear that Government has authorised further research into DNA testing of faeces and hairs outside badger setts that could determine whether the badgers in that sett were infected with Bovine Tuberculosis. This is a terrible disease for livestock farmers, the latest published statistics show that 29,832 cattle were compulsorily slaughtered in the year up to October 2015 after failing the TB skin test for tuberculosis. Most of those cattle had contracted the disease from infected badgers. At the present time control of infected badgers is by authorised shooting in specific cull zones, but further research is being carried out into a suitable gas which could then be used to kill the badgers in the infected setts in a humane way, so that the disease could be controlled without harm to any healthy badgers.

The Clarion – *Summer 2016*

Open Farm Sunday. Roofs of Farm Buildings. Risk Analysis HSC/06/55. Different Types of Asbestos.

Open Farm Sunday

We look forward to welcoming readers of Clarion V to Kensham Farm between 2:00pm and 5:00pm on Sunday when we will be open for 'Open Farm Sunday' – this is a national initiative run by LEAF ('Linking Environment and Farming') when around 400 farmers will welcome the general public, especially youngsters, onto their farms to see how food sold in the shops is produced on the farms.

We generally alternate this event with Laceys Family Farm at Bolter End, so this year it is our turn – but there should be some of Laceys' young cattle or calves on display, as well as our farm machinery and tractor rides around our arable fields where we grow wheat for bread making.

Cadmore End School Parent Teacher Association will serve teas in our farm machinery workshop. Visitors will be able to see round some of our farm buildings, including our newest grain dryer and store for 3,000 tonnes which was built in 2013. The Svegma grain dryer, made in Sweden, can dry the grain at 46 tonnes per hour, whereas when we installed the last new continuous flow grain dryer in 1961 it was rated at 1.25 tonnes per hour – this is the extent to which the pace of farm production has increased during the past 55 years.

Roofs of Farm Buildings

Suitable farm buildings are an essential part of farm production, whether in the form of stockyards for cattle, lambing pens for sheep, piggeries, hen houses or just storage buildings and machinery repair workshops – and they must all have roofs to keep the inside dry. Some visitors may look at the roofs of our farm buildings, and if anyone should have any worries about health risks from asbestos cement on the roofs I hope that they will discuss their concerns with me.

Health & Safety Commission – Risk Analysis HSC/06/55

The histogram bar chart below was published in 2006 in a risk analysis paper by the UK Health & Safety Commission entitled HSC/06/55 – available to view on the Government HSE website. It shows the comparison of risks to health from different materials which contain 'asbestos'. The reason for these differences in risk is that 'asbestos' is a generic term, which includes the dangerous amphibole forms such as 'Blue' and 'Brown' asbestos, as well as Chrysotile, often called 'White Asbestos'- which has never caused any ill health, and is the type of asbestos used for the manufacture of asbestos cement.

The tall bar in the chart shows the real risk to health caused by inhaling the sharp fibres, like tiny glass needles, from the amphibole form of asbestos which was used before the 1970s in spray applications of insulation. At that time Governments throughout the world realised the danger to workers who worked in factories using blue and brown asbestos, where often the air which they breathed was full of those harmful asbestos fibres. That was the reason why all future mining or use of blue and brown asbestos was forbidden by law.

The chart also shows some risk to those factory workers who used to work for prolonged periods manufacturing Asbestos Insulation Board ('AIB') which was used before 1999 to protect parts of buildings from fire risk. The negligible risk to health from manufactured products containing asbestos, such as flooring, car battery cases and asbestos cement, is shown clearly. The International Chrysotile Association of Canada make the point that the only recorded deaths caused by work with asbestos cement have been falls, when roof repairs have been attempted without using crawling boards. Such falls are an even greater risk with new mineral cement roofing sheets than with work on roofs which had been built with asbestos cement before its use was banned in the UK.

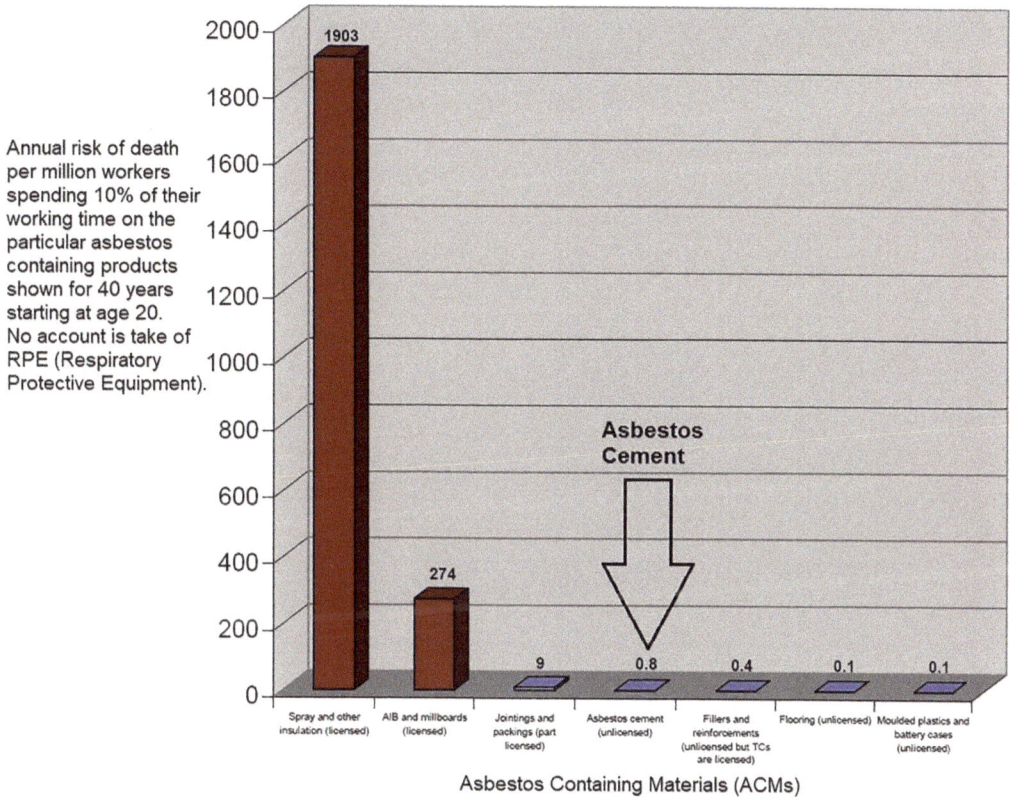

Risk Analysis from the Health & Safety Commission paper HSC/06/55, showing annual risk of death per million workers from working with different asbestos containing materials

Annual risk of death per million workers spending 10% of their working time on the particular asbestos containing products shown for 40 years starting at age 20. No account is take of RPE (Respiratory Protective Equipment).

2000
1800
1600
1400
1200
1000
800
600
400
200
0

1903

274

9   0.8   0.4   0.1   0.1

Asbestos Cement

Spray and other insulation (licensed) | AIB and millboards (licensed) | Jointings and packings (part licensed) | Asbestos cement (unlicensed) | Fillers and reinforcements (unlicensed but TCs are licensed) | Flooring (unlicensed) | Moulded plastics and battery cases (unlicensed)

Asbestos Containing Materials (ACMs)

Histogram bar chart from the Health & Safety Commission paper, HSC/06/55, shown, annual risk of death per million workers who have worked with different asbestos containing products for 40 years. Less than 1 worker per million working with asbestos cement could be expected to die, whereas 1,903 workers involved with spray insulation (now banned) could have been expected to die.

Different Types of Asbestos

The reason for the differences in risk to health from the different types of asbestos can be explained easily. All forms of asbestos are naturally occurring fibrous silicates, which are mined in countries such as Canada, Brazil and Russia. The blue and brown forms are iron silicates, containing sharp needle like fibres which are insoluble in the acid of the human lung. These harmful fibres, when inhaled deep into the lungs, will remain there - sometimes for forty or fifty years, before finally penetrating the lung and irritating the mesothelium which surrounds it, thus causing mesothelioma or lung cancer.

But the white form of asbestos, chrysotile, is a magnesium silicate in which the fibres are soluble in acid, with a texture more like soft wool. So if any fibres from asbestos cement should unfortunately be inhaled deep into the lung they will be dissolved within about two weeks by the natural acid in the lung. Chrysotile fibres have never been shown to have caused mesothelioma.

That is why asbestos cement is a really good and safe roofing material, which farmers in USA, Canada, Russia, and most other countries in the world, can still purchase and use without any restrictions. As a building material it is more durable and better than the modern 'Mineral Cement' roofing sheets which have had to be used on UK farm buildings – ever since the sale of products containing any type of asbestos was banned in Europe and the UK, and the EU Waste Framework Directive failed to differentiate between the amphibole and the chrysotile forms of asbestos.

This further histogram chart was published in July 2006 by HSE as annex to Paper HSC/06/55, comparing Asbestos Cement and Textured Coatings to other public risks.

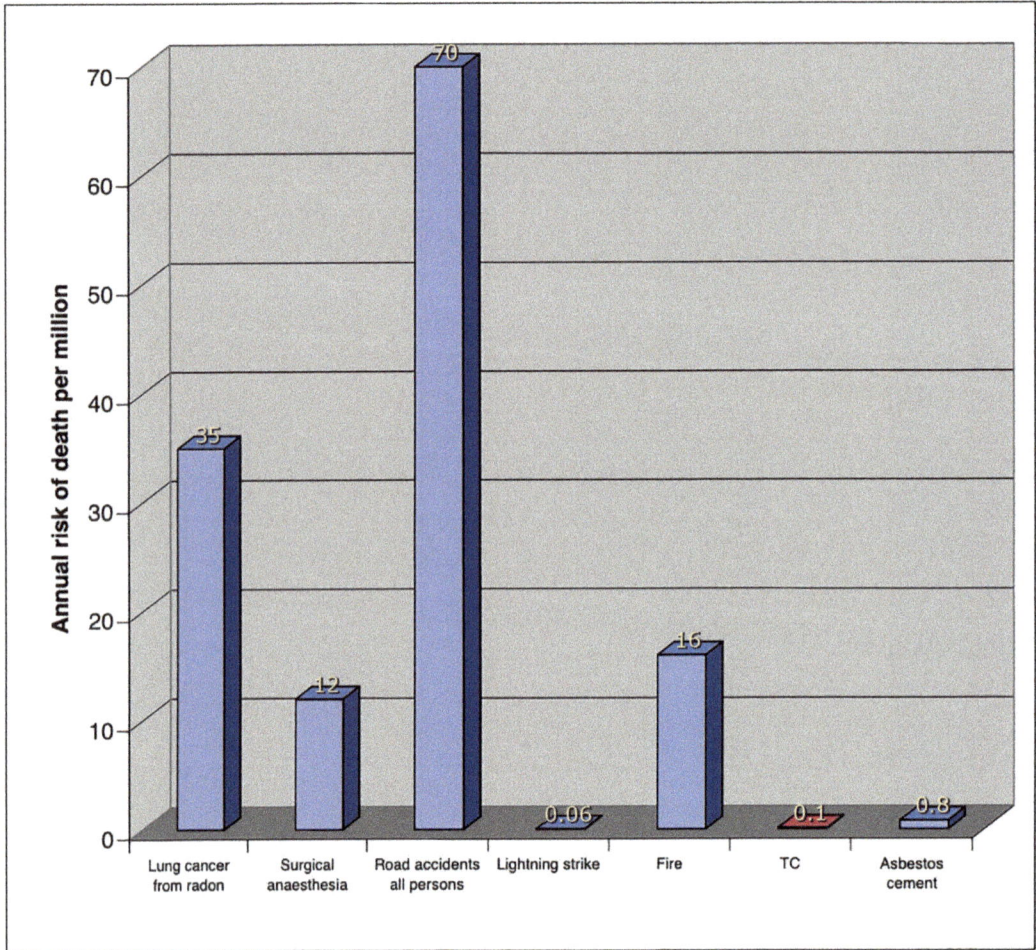

Comparisons of textured coating and asbestos cement product group annual risk of death per million to other public risks. (asbestos risk based on 10% of time actively removing ACMs from age 20 for 40 years with limited controls and no RPE).

The Clarion – *Autumn 2016*

'Brexit' Referendum. NFU Key Aspects for Regulation of British Farms. Harvest 2016.

The 'Brexit' Referendum

Three weeks after publication of the last issue of Clarion V the 'Brexit' Referendum took place, with a decision for Britain to leave the EU, so that our farming policy in the future will be formulated and controlled by British politicians in Westminster.

The head of the National Trust is reported in the national press as saying that she wants to return the countryside to how it was "in our parents' and grandparents' generations" under a new post-Brexit British farming policy. She thinks that food production does not matter, and that the land should be managed in a nature friendly way to benefit farmland birds, wildflowers, bees and butterflies.

Furthermore, she feels that "water meadows and meandering rivers will help prevent the flooding of our towns" - but unfortunately she makes no mention of the extensive flooding in the Somerset Levels which we reported in the Spring 2014 issue of Clarion V. At that time the floodwater could not drain out to the sea quickly enough because the River Parrett had not been correctly dredged or maintained in recent years - until emergency work had to be carried out to alleviate the flooding. And the reason why the Environment Agency had failed to dredge it correctly was the result of its misguided policy of treating the habitat and well-being of water voles in the riverbank as being more important than protecting houses from flooding.

John Deere 5680i Hillmaster combine harvester on sloping ground. The header will follow the contour, whereas the cab and threshing mechanism will remain horizontal.

NFU Key Aspects for Regulation of British Farms

So farming policy has to be finely balanced to grow food for human consumption, as well as protecting the landscape and wildlife. The NFU has set out seven key aspects of the way in which we hope that our British politicians will regulate UK farming. These are: -

1. Build the nation's food security by improving the competitiveness, profitability and productivity of UK agriculture *(Government to treat food production as important)*

2. Give British farmers the best possible access to markets inside and outside the EU *(for sale of farm produce)*

3. Ensure that agri-food imports *(such as bacon, pork, lamb, beef and poultry meat)* meet the same high standards adhered to by British farmers

4. Introduce a domestic agricultural policy that will not put British farmers at a competitive disadvantage *(as compared to food produced overseas)*

5. Ensure British farmers and growers have sufficient supplies of labour, *(many seasonal workers in the vegetable growing sector come from Eastern Europe)*

6. Strengthen farming's environmental role, allowing all farmers to care for the countryside, wildlife and mitigate climate change *(as a by-product of farming)*

7. Implement policies that are science and evidence-based to create a better regulatory environment for British farmers *(no unnecessary regulations)*

Harvest 2016

At Kensham Farms our harvest started this year on 30th July, with a crop of oil seed rape on the Grove Farm field next to Booker Aerodrome, followed by harvesting winter oats at Pyatts Farm, shown in the photo. Our combine harvester is a John Deere S680i Hillmaster combine. One of the photos shows how the cutter head, which is 9 metres wide, can follow the slope of the land and yet the cab and the thrashing mechanism of the combine remains level. The mechanism to achieve this tilt is by the link arms for the driving wheels pushing down one wheel and raising up the other - in just the same way that a dog can run along the contour line of a hill without falling over, by lowering one leg relative to the other.

The combine harvester has an 11 litre diesel engine, consuming 70 to 80 litres per hour, it develops 565 hp. It cuts the standing crop, then thrashes and cleans the grain at a rate of 45 to 50 tonnes per hour, just occasionally reaching one tonne per minute - in a full day we can harvest around 100 acres. It is fitted with GPS steering and has all the latest computer technology - so that after harvesting each field it will transmit a map of the field, showing yield and moisture content of the grain, by radio link to the farm office.

Wheat harvesting, at an average rate of 45-50 tonnes of grain harvested per hour.

The Clarion – *Winter 2016*

Currency Rates of Exchange. Mustard Plants as Game Cover. Importance of a Good Seedbed.

Currency Rates of Exchange

Since the last issue of Clarion V the national press has been full of reports of the falling value of the £ sterling against other currencies, with some of those reports regarding this as a disaster. But as far as the farming industry is concerned the lower exchange rate has been good news, in that UK farm exports, such as grain and cheese, now return a better price in £ sterling to the farms where the raw ingredients for this farm produce were grown. And since those exports set the tone of the market, similar food products used for UK home consumption are also sold for a better price than pre-Brexit, although still not at the satisfactory prices that we received about three years ago. One downside will be an increase in the cost of imports, such as diesel oil and farm machinery imported from the USA or Europe.

Horsche Joker cultivator.

We have recently bought a new cultivator, a German Horsch Joker disc cultivator, shown in the photos. Initially our farm machinery agent supplied a demonstration model without charge – we used this for several days and found it to be better and heavier than our earlier disc cultivator, and able to work with a more consistent depth, for cultivating fields where there is trash which has to be broken up and buried. We have also found it excellent for burying and incorporating the Thames Water bi-recycling cake organic manure which we use – the old-fashioned name for this product was sewage sludge.

Mustard Plants as Game Cover

The photo was taken on 20th October on one of the West Wycombe Estate fields by Bullocks Farm Lane which we farm. We had seeded mustard into the previous growing crop of wheat, so that after harvest it would make good cover for the Estate partridge shoot in September and early October, and then provide useful organic matter for incorporation into the earth to benefit the next crop. When the mustard cover crop had completed its job for the Estate shoot, we mowed the green mustard plants, which were standing around three or four feet tall, and buried them by cultivating with the Horsch Joker – the photos show the second pass of the Joker cultivator burying the mustard plants and working the soil down to a good seedbed for seeding the next crop of wheat, which should be completed before the end of October. So the Joker has made a good job of preparing the field for the next crop but its cost, at £35,500 for this ex-demonstration model, was no joke at all.

Importance of a Good Seedbed

A good seedbed is just as important on the farm scale as for an allotment or vegetable garden, in that the aim is to eliminate all trash or weeds, and then work down any clods of earth to prepare clean and friable soil in which the new seeds can grow. Moisture in the soil is very important, in that a seedbed which is wet or sticky will not lead to good germination and satisfactory early growth of the new seedlings. This year the largely dry September and October have provided excellent conditions for Autumn cultivations, leading to the strong young wheat seedlings shown in the smaller photo, which was also taken on 20th October, on our Kensham End field only 16 days after seeding. As a general rule if a field at seeding time is dry enough to walk over wearing shoes then the seedbed will be good, but if it is so wet and sticky that it cakes onto gumboots then the new crop is not likely to establish well. If early establishment of the crop is poor, the final result at the next harvest is likely also to be disappointing – as it was after the wet Autumn of 2012.

Cultivation Methods

The different cultivation tools on a farm are similar to the different clubs in a golfer's bag, with some clubs suitable for long shots, others to get out of bunkers and others for the fine tuning on the putting green. So on the farm we have a 7 furrow plough, suitable for turning the soil right over to bury stubble or weeds, but ploughing is more expensive and uses more fuel than our heavy Sumo cultivator, with its discs and sub-soil tines to loosen the soil below normal working depth, which is suitable for the first cultivation of clean and weed free stubbles. We also use other cultivators like the Joker described above, or our Rexius roller and leveller to follow the plough, and then lighter harrows, sometimes with spring tines. The final work of smoothing the seedbed prior to seeding with an 8 metre width seed drill, is pressing with Cambridge ring rollers to provide the friable and firm soils conditions, with no clods or cavities, which should lead to good crop establishment.

Winter wheat seedlings photographed 16 days after seeding.

The Clarion – *Spring 2017*

Methods of Farm Support. Decline of British Farming in the 19th Century. German U-boats in 1914. World War II Food Rationing. Agriculture Act 1947. BCA Student Visit. Oxford Diocesan Plough Wednesday.

Methods of Farm Support

I am sometimes asked what farm subsidies are all about – a subject which will be greatly influenced by the Brexit vote for Britain to come out of EU control. This was debated at the Oxford Farm Conference in January, and also in the very same week at the alternative 'Oxford Real Farming Conference' where it was suggested by an environmentalist that *'supporting farmers to produce food is morally wrong'*.

This has prompted me to take a close look at the issue of EU support for the UK Farming Industry, which in recent years has equated to roughly the same value as total net profit for British farms. Thus, if there had been no direct farm support (either from the EU or from the UK Government), nearly all food production in Britain would have closed down. It is only incidental that many farms would probably have continued in business, with farmers earning a living from their farms by providing horse riding facilities, B&B accommodation, farm shops, turning farm buildings into workshops or factories, field sports, providing 'nature trails' or Go-Cart tracks, and charging the general public for outdoor recreation.

As a generalisation all people have to eat to keep alive (and plenty of us, including the author, enjoy eating) so they have to get food from somewhere.

Decline of British Farming in the 19th Century

In Britain we had a prime example of what happens if food supply is left to the global free market in the second half of the 19th century. Food production in Britain declined catastrophically following the Repeal of the Corn Laws – these had been introduced by the Importation Act 1815 were subsequently repealed by the Importation Act 1846, at a time when wheat grown cheaply on the newly farmed Prairies of the USA was imported into Britain at a cost that was below cost of production of wheat grown on the smaller fields of Britain. This decline of home-grown food production was accentuated when larger ships were developed, thus further reducing the cost of freight, so that British farming reached such a low point by the 1860s that there was extreme poverty in all rural areas of Britain (other than for the aristocracy), and many farms became derelict in that pre-First World War period.

German U-boats in 1914

Following declaration of war in 1914 a primary objective of Germany was to win the war by causing starvation in Britain, hoping that this would lead to its capitulation. At that time German U-Boats were successful in sinking many British merchant ships in the convoys bringing the food to the UK, and Britain nearly starved – so farmers in Britain were urged to recommence food production, and started to receive adequate payment for producing food at a price that enabled them to stay in business.

After World War I the UK Government could not control worldwide depression, and yet it had a responsibility to avoid British folk from starving, so by the end of the 1920s it was politic to get the cheapest food possible from somewhere, whether it was from home farms, or from overseas farms with cheaper land, labour and food production resources. Thus British farming declined again, arable crop production nearly ceased, and those farmers who survived largely turned to 'dog and stick' farming – a system in which they did not cultivate anything, but just kept some livestock for beef or lamb production on the grass meadows.

World War II Food Rationing

Then war was declared yet again, this time World War II, when the Germans tried to starve out Britain by sinking the ships in the food supply convoys with torpedoes from U-Boats. UK Government launched the critically important and successful 'Dig for Victory' campaign (in which the author's grandfather took on an allotment in addition to his own vegetable garden, and the author's father dug up a grass tennis court to grow extra vegetables on it) and it was as successful as the front line troops in saving our nation - by growing enough food, in conjunction with food rationing, to feed us all.

The Agriculture Act 1947

Then, just after World War II ended, the Labour Minister of Agriculture, Tom Williams, brought in the very sensible Agriculture Act 1947 – declaring to Parliament its aim as being:

> *"to promote a healthy and efficient agriculture capable of producing that part of the nation's food which is required from home sources at the lowest price consistent with the provision of adequate remuneration and decent living conditions for farmers and workers, with a reasonable return on capital invested".*

The author's final comment on this subject of Government support for growing food, is that direct farm support (funded in recent years by the EU for Britain, but with most other countries in the world, including the USA, having their own support systems) enables ordinary citizens to buy food, with a certainty of availability at price that is less than cost of production. Direct farm support for food production is therefore really a subsidy that provides cheap food for everyone. There is also the further advantage that the countryside and wildlife can continue to be looked after by farmers as a by-product of the production of this food.

BCA Students learning about Grain Production

It is difficult to take an interesting photo of a debate, and harder still to take an interesting photo of a subsidy. So, these two photos were both taken on recent study days on farming and grain production. The first is of Charlie in our farm workshop at Kensham Farm explaining to students at the Berkshire College of Agriculture (BCA) how to dry and store grain so as to keep it in top condition from harvest time until it is sold, sometimes as late as May or June in the following year, until it is required by flour millers for making bread.

Charlie explaining to BCA students how to dry and store grain.

Oxford Diocesan Plough Wednesday

The second photo was taken at the flour mill of FWP Matthews Ltd at Chipping Norton where this mill, still managed by the Matthews family, has been milling wheat, much of it from local farms, since 1912. The event was the Oxford Diocesan Plough Wednesday – a training day on 11th January for clergy and others involved in rural ministry. All of us delegates had to put on the white overalls, high visibility jacket and blue hair covering seen in the photo as part of the routine hygiene precautions at the mill, where the production manager was explaining to us the process involved in milling wheat to make flour.

Oxford Diocesan Plough Wednesday Training day visit to the Flour Mill of FWP Matthews Ltd at Chipping Norton.

The Clarion – *Summer 2017*

Rainfall Records. Mains Water Supply. Johne's Disease. Portakabin Crane Hire.

Rainfall Records

Our records over the past 10 years show that our annual average rainfall at Kensham Farm has been 859 mm (33.8 inches), but in April 2017 the rainfall was only 5.5 mm. This had made it an unusually dry Spring - but not quite the driest, in that ten years ago in April 2007 the rainfall at Kensham Farm was only 3 mm. However, the crops were not short of water in that year, since the average for the following three-month period of May to July 2007 had been 133 mm (over 5 inches) rainfall in each month.

Those statistics have also reminded me that I have now been writing a few thoughts about the farming scene each quarter for The Clarion for 10 years, so this issue starts my eleventh year. I have often commented that good weather and soil conditions are the most important factors in making a good seedbed, and without a good seedbed crops will be off to a poor start, from which they seldom recover. This season with dry conditions in both in September and October for Autumn seeding, and in March for Spring seeding, we were able to make good seedbeds, so that at seeding time it was possible to walk over the fields in shoes rather than in boots, with the soil, just like a good gardening soil, in a nice friable condition to handle – not sticky or wet. But good germination has to be followed by sufficient rain to keep the crop growing.

Mains Water Supply

We have had an unwelcome reminder of the importance of water, for farming as well as for everyday life, in February this year, when we received a Thames Water bill for over £10,000 for water for 3 months, whereas our normal quarterly bill is around £400 for 3 months. Our investigation showed that the estimated bill was quite wrong, but the actual consumption had been £2,000 of water, the excess over our normal consumption having been caused by leaks.

At Kensham Farm there was no mains water until 1948. Before that time all the rainwater from the gutters of the farmhouse had been saved in an underground brick-built chamber, rather like a well in construction, but only 7 or 8 feet deep and designed just to save water, rather than to go right down to reach the natural ground water level. Then in the kitchen the water was pumped up with a semi-rotary hand pump for use for cooking and washing. And at that time no water was wasted flushing the loo, since the sanitary arrangement was just a brick built shed in the garden, with a one-hole earth closet.

Installation by farm staff of new water main for the farm buildings and paddocks.

Johne's Disease

But for the cattle which were kept by our predecessors, Jack and Frank Bird (whose father had started farming here towards the end of the 1800s), the health aspects of pond water were not so good, since the cattle had to drink out of the pond on Cadmore End Common, just in front of the farmhouse. And that pond had become infected with Johne's Disease (Paratuberculosis), a nasty chronic enteritis caused by this mycobacterium from which there is no prospect of recovery or treatment, so each year the Bird Brothers lost two or three of their cattle which had suffered from the disease so badly as to lose weight until all the ribs on their body could be seen, followed by death.

When the new piped water system was laid in 1948 galvanised water troughs were set up in all the fields that were used for grazing the cattle, and after that there were no new cases of Johne's Disease. This proved that the infection had been in the pond, made worse by the cattle fouling in the very pond which provided their only water supply. But nothing lasts forever, and by 1974 the galvanised water pipes had started to rust, perhaps the quality of the zinc galvanising just after World War II had not been as good as nowadays. So at that time we had to replace all the underground galvanised iron water pipes with the latest recommended type of water pipe, black imperial size alkathene. We did have subsequent difficulties making any extensions or repairs to the system, since manufacture of the black imperial size ceased when it was replaced by the newer blue alkathene, needing special couplings to convert from imperial to metric.

When we investigated the cause of our excessive water consumption it transpired that the original black imperial size alkathene becomes brittle with age and can then crack – leading to serious underground leakage which is difficult to spot. So for a second time we have had to replace many of our underground water pipes, this time using the blue metric sized alkathene which is supposed to have a longer life – but time will tell, and perhaps my grandsons will have to replace all the pipes again in the years to come. The photo shows us backfilling the trench after laying a pair of new main pipes on Cadmore End Common, one to the farmhouse and another bigger pipe for all the troughs and standpipes on the farm. The pond is on the right – it is now a haven for ducks and wildlife, and we hope that it no longer harbours any of the Johne's Disease infection which had been in its water before the days of mains piped water.

Portakabin, Crane Hire

The other photo was taken when we had a second hand 4 tonne Portakabin installed for one of our farm workshops. The difficulty was the telephone lines that can be seen in the photo, in that the delivery lorry was one side of the lines, and the site for the Portakabin was the other side. However, Lane End is a good village, and we did not have to go any further afield than Meakes Blacksmiths Works on Ditchfield Common to hire a crane – with a driver for which such a task presented no difficulty, just part of the day's work.

Craning in a portacabin for use as an office by Sprayfinishes, licencees of one of the Kensham Farms workshop units in former poultry houses.

The Clarion – *Autumn 2017*

Politics and Shortcomings of the CAP. Start of Harvest. Farming Calendar.

Politics and Shortcomings of the Common Agricultural Policy

For the fortunes of farmers political factors have historically always mattered more than those vagaries of weather and the forces of nature which might be expected to be the main concerns of farmers. Since the last issue of Clarion V a new Minister for Defra has been appointed, none other than Michael Gove who at one time was not the favourite politician of school teachers.

But looking on the bright side, it is likely that UK farming will no longer be controlled by the one-size-fits-all policies of the EU, which so often failed to recognise the merit of the latest scientific discoveries affecting food production. And Michael Gove has recognised that the EU Common Agricultural Policy (CAP) and the Common Fisheries Policy both had serious shortcomings which he hopes to put right in the years to come, but he does not expect an immediate solution to be possible.

The acronym 'Defra' stands for the Department for the Environment, Food and Rural Affairs, a title which farmers have always felt was incomplete in that it makes no mention of agriculture or farming. But the new Minister has recognised that the first reason for Government policy to continue to support farmers is the high quality of the food which they produce – a really good starting point.

Start of Harvest

So with that in mind, and turning to this year's cereal crop harvest, we made a much earlier start than usual, with the winter barley harvest starting on 8th July. The photo shows harvest progress on 9th July on our Top Plain field, on the left of the B482 road into Stokenchurch. This is a large field which yielded 335 tonnes of grain off its 83 acres, an average of just over 4 tonnes per acre. The variety was Bazooka, a newly developed high yielding hybrid variety, producing grains suitable for livestock feeding – mainly pigs, poultry and cattle.

Growing cereal crops has become far more sophisticated over the years since we started farming here at Michaelmas 1955. At that time we were pleased when a field of barley yielded more than a tonne to the acre, but now we expect the yield to be three or four tonnes per acre.

The main improvements have been in plant breeding, in the development of fungicides to control such plant diseases as mildew, yellow or brown rust, septoria, or rhynchosporium. Other improvements have been better herbicide weed killers, and the development of growth regulators which make the cereal plant grow shorter but with a thicker stem and stronger root system.

Taking the barley crop shown in the photo on Top Plain as an example, and in particular the timings of the different cultivations and treatments which it received, we can set out these treatments by month as: -

Offloading the combine on the run. The tractor with trailer is straddling the row of straw, which will be bailed within the next day or two.

September 2016
Glyphosate (Roundup) to kill any weeds in the seedbed, and an insecticide to kill any aphids. The seed was planted on 22nd September with our 8 metre wide seed drill, then before the crop had grown up through the soil we sprayed the soil with a mix of two different herbicides. These pre-emergence sprays will kill any weeds that try to poke up through the treated crust of the soil, but do not harm the barley crop as it emerges.

October 2016
The first fertiliser, Muriate of Potash, was spread in October. This particular field did not need any phosphate fertilizer this season, since it still benefitted from the phosphate rich sewage sludge which had been spread three years earlier.

March 2017
The crop was sprayed with a 'tank mix' of two different types of fungicide, plus a grass weed herbicide, plus two different types of growth regulator. Then three applications of nitrogen fertilizer were spread, two of granular urea and one of ammonium sulphate.

April 2017
In early April the crop was sprayed again with a tank mix of two different fungicides plus

two types of growth regulator, then a third spray treatment of growth regulator was applied at the end of April. This growth regulator was to ensure that the crop remains standing upright, without lodging down onto the ground.

May 2017

Two more different types of fungicide were sprayed onto the crop leaf on 9th May. Some types of fungicide are only active for around three weeks, so they need to be re-applied to keep the growing plant and fresh leaves free from disease.

July 2017

All the crop treatments shown above are recorded on our computer crop record for the field. No treatments were carried out in June, and harvesting started on 8th July.

Immediately after harvest the straw, which had been left in swathes behind the combine harvester, was baled and carted off by the Lacey family for use in their cattle yards during 2017/18 winter. Following that, sewage sludge (marketed under the more polite name of 'Digested Cake') was spread to provide the necessary phosphate for the next three years of cropping. Finally, the field was then cultivated and rolled to make a seedbed in which we hope that any weed seeds will germinate – so that they can be killed off with glyphosate, prior to seeding oil seed rape for harvesting next year, probably in July 2018.

Completion of 2017 harvest, From the left:- Alex, Charlie, Bryan, Nigel Rogers, driver of the Kensham Farm combine for 57 years, and Angus.

The Clarion – *Winter 2017*

Old Year Quarter Days. Stability of British Food Production. RSA Food, Farming and Countryside Commission. Royal South Bucks Agricultural Association.

The Quarter Days of the Year

Michaelmas Day, the 29th of September, was one of the Quarter Days with origins as religious festivals. The others were Christmas on 25th December, Lady Day on 25th March and then Midsummer Day on 24th June. On the Quarter Days in ancient times debts and unresolved lawsuits were not allowed to linger on – a reckoning had to be made and publicly recorded.

But for farming Michaelmas Day had far more significance, in that it was the day in each year when Mop Fairs were held, sometimes called Hiring Fairs, when both male and female agricultural servants would gather in order to bargain with prospective employers, and hopefully secure a position for the coming year. They had been held since 1351 when labour was short after the Black Death and continued until a hundred years ago, when such fairs were still being held by the Guildhall in High Wycombe. Then the Corn Production Act of 1917 was enacted, for the purpose of ensuring that there was sufficient food available to feed the nation during the First World War. It achieved this by guaranteeing a minimum price for wheat and oats, and it set up the Agricultural Wages Board, with a minimum wage for agricultural workers, thereby ensuring stability for both the farmers and the workers on the farms.

Stability of British Food Production

With the recent Brexit vote, once again Government is having to give thought to those objectives, to ensure stability of food production in this country, rather than just relying on imported food. The last in depth review was in 2001 when Defra, the Department for Environment, Food and Rural Affairs, set up the Policy Commission on the Future of Farming and Food, under the chairmanship of Sir Donald Curry – but its terms of reference had to be consistent with the Government's aims for CAP Reform, enlargement of the EU and increased trade liberalisation for food imports from other countries.

The Royal Society for the Encouragement of Arts, Manufactures and Commerce (the RSA) has recently set up a similar new review - the Food, Farming and Countryside Commission, most appropriately while the UK negotiates its future relationship with the European Union. But the big difference this time is that the Commission's work is not being funded by Defra or any other Government department, it is being funded by the Esmee Fairburn Foundation, a charity which provides major grants for projects for the national good, and will be administered by the RSA.

So this new Commission has no obligation to Government or any political party - its terms of reference are to prepare the Commission's Report within two years, and to come up with recommendations that should secure the future of food, farming and the countryside in Britain. It was good that at the launch event the Chairman, Sir Ian Cheshire, finished his answer to two of the questions from the floor with the following words: -

> *".....we cannot have farming without farmers and the rural economy in the broader sense of the word and I absolutely take the point that will include the subsidies and payments. The nature of this model is going to be multiple and complicated, but we have to have a genuine long term future for farmers. There is absolutely no doubt in my mind about it. That relates to the final point on food security"*

Royal South Bucks Agricultural Association

Concerning local events, rather than matters of national importance, on the first Wednesday in October the annual Ploughing Match of the Royal South Bucks Agricultural Association took place - this year in Lord Parmoor's field, farmed by the Connell family, between the Frieth old horse pond crossroads and Finnamore Wood. RSBAA is a historic association which had its roots in the research carried out by the Revd. St. John Priest published in 1813 for the Board of Agriculture to formulate policies *'to improve agricultural practices and economics'*. This led to the association being formed in 1833 when King William IV was on the throne, before Queen Victoria.

This year was the 173rd RSBAA Annual Ploughing Match, at which six teams of horses were ploughing, as well as modern tractors and ploughs, and several ploughs and tractors dating from around the 1950s. One of the photos shows a horse team at work on Show Day, the pair of horses pulling a single furrow plough. That type of ploughing was hard work for the ploughman, who had to walk in the furrow behind the plough all day, and also required great skill in setting up opening ridges and closing furrows, since the plough always turned the soil over in the same direction, towards the right, whether it was going up the field or down it.

The modern six furrow reversible plough shown in the other photo, taken at Bullocks Farm, Wheeler End in November, not only covers more ground in a day than the horse plough would cover in a couple of weeks, but has the advantage of not needing opening and closing furrows. This is achieved by using the left hand mouldboards shown at work in the photo when going in one direction, then at the end of the field the tractor hydraulics twist the plough over so that the right hand mouldboards (which are up in the air in the photo) are used on the return run along the field.

Following the RSBAA ploughing match there was a three course lunch in a marquee for 470 members and guests, at which it was my great pleasure to present Long Service Awards of 55 years at Kensham Farm for Nigel Rogers, and 18 years for his son Paul Rogers. So I am pleased to report that the heading for this column in the Spring 2014 issue of Clarion V that 'Farm Foreman Retires after 52 years' was not in fact accurate, in that Nigel continues to drive our combine harvester. Records of RSBAA show that the longest Long Service Award ever was to James Lesley in 1882, when he received an award of 15 shillings in recognition of 58 years of service – so there is a great incentive for Nigel and myself to continue for a few more years.

One of the horse teams ploughing with a single furrow plough at the Annual Ploughing Match of the Royal South Bucks Agricultural Association.

Modern six furrow reversible plough shown with the left hand mould boards in work.

The Clarion – *Spring 2018*

Soil Temperature. Fungicides and other Crop Treatments.

Soil Temperature

March is normally the month of the year when field work starts after the winter months, since in the months from November to February soil temperatures are too low for growth, and the soil condition is generally too wet to take the weight of a tractor. On our Kensham Farms fields our main crop is winter wheat, but this year we are also growing barley, both winter and spring seeded, and triticale, which is a cross between rye and wheat. We are also growing oilseed rape, which was seeded in August 2017 on Top Plain, the large field on the left of the B482 road between Cadmore End and Stokenchurch. This will look bright yellow in April or May 2018, before the small dark seeds with high oil content are ready for harvesting in July 2018.

We try to make full use of modern crop protection chemicals and artificial fertiliser so as to maximise our production of grain – an objective which is only possible if the crop is kept healthy, well nourished and free from disease during the growing period. Research scientists have developed crop protection products in recent years which are far more effective than those which were available when we started our farming business here at Kensham Farm in 1955.

Fungicides

One of the greatest improvements has been in the field of fungicides. Sixty years ago no effective products were available to control such diseases as mildew on the leaves of growing crops. The leaf of a growing plant of wheat or any other cereal should be kept in a healthy blemish free condition for as long as possible, so that it can photosynthesise – this is the process whereby sunlight shining on the leaf of the plant enables the green chlorophyll pigment in it to convert the light energy into chemical energy, forming carbohydrates such as sugars from the carbon dioxide in the air and water from rainfall. Most of the oxygen which we breathe is a by-product of this conversion of carbon dioxide into carbohydrates by the plant when it has sufficient sunlight and moisture.

This is the reason why it is so essential for the plant leaf to be healthy – if it is covered in mildew or plant disease, so that the leaf is no longer a healthy green colour, the plant will cease to grow and provide sufficient food that can be sold for human consumption after harvest. In order to keep the plant's leaf area in this healthy condition we use a variety of different fungicides, each one formulated to be most suitable for the stage of growth that the plant has reached.

If we take the example of growing a crop of wheat, the grain from which may eventually be suitable to mill into flour for bread making, we can look at the most important treatments which the plant of wheat will receive throughout its life. These crop treatments are: -

Spring seeding of wheat into a clean and well prepared seedbed.

Preparation of the Seedbed. All gardeners will know the importance of preparing a good seed-bed, which is free of weeds. This can be achieved partly by cultivations, and partly by using the total weed killer glyphosate (generally sold under the trade name 'Roundup'). The soil must also be in nice friable condition with plenty of humus, particularly avoiding an excess of moisture which might cause the wet soil to become compressed and waterlogged.

Seeding. The seeds for the crop are carefully selected from the best quality seeds from the previous harvest. These seeds will be coated with a seed dressing which will protect the growing seedlings from such seed born diseases as septoria, loose smut, seedling blight or bunt.

Fertilisers. The plant will not grow satisfactorily if it is short of plant food, the principal nutrients being Nitrogen for leaf growth, Phosphates for the root system and Potash for general plant health. The general aim on farms is to use sufficient Phosphate and Potash fertilizer to maintain the soil at a satisfactory status, verified by soil analysis. But Nitrogen is different, in that it is water soluble and does not last in the soil - it has to be applied during the growing season to suit that particular crop's requirements. These plant nutrients will only benefit the growing plant if the soil is neither too acid nor too alkaline - the natural acidity of some soils can be neutralised with a dressing of lime. In this area those fields that are chalky never need additional lime, but the fields with clay soil may need a dressing of lime, perhaps thee tonnes to the acre, every third year.

Herbicides. Pre-emergence weed killers have been developed and are often used – these can be sprayed onto the soil before the crop emerges, and will kill any weed seedlings as they emerge through the soil crust, but the wheat seedlings will not be affected by this pre-emergence spray residue on the treated soil crust. However as the crop grows some weeds such as chickweed, charlock or poppies are likely to appear in it, and must be killed to prevent them competing for plant nutrients and sunlight with the growing crop. These weeds should be killed with a suitable hormone weed killer, probably by spraying the field in April or May, which will kill the weed without adversely affecting the crop.

Growth Regulators. Scientists have developed growth regulators which will cause the growing crop to grow shorter than its natural height, but the stem and the root of the growing crop will become stronger than untreated wheat. Treatment with growth regulator is the reason why nowadays one seldom sees crops which have lodged, those disastrous crops that had gone down flat during a wet growing season and turned harvest into a salvage operation.

Fungicides. The fungicides are the newly developed crop protection products which keep the plant leaf healthy. The aim with fungicide is always to anticipate trouble and apply the preventative fungicide spray before the disease has had time to spoil the foliage. It is not unusual for a crop to be treated with five different fungicides, each specific for a particular plant disease that may be anticipated at a particular time of the growing season.

Insecticides. Gardeners will be familiar with the problem of controlling aphids on vegetable crops like broad beans, or on flowers such as roses. Aphids can cause similar problems in cereal crops, such as barley yellow dwarf virus, which must be controlled by spraying suitable insecticides – sometimes earlier in the year to control the aphid's life cycle if trouble is anticipated, but sometimes when the aphids are actually on the crop. Many crops of wheat never have to be sprayed with insecticides - these are always used as sparingly as possible to avoid harming bees or beneficial insects.

All of these crop treatments have the objective of growing full crops on the farms to feed the world's expanding population of people. Most farmers feel that politicians who fail to recognise the advances made by scientists in the field of crop protection treatments have no recollection of the poor crops and low production of earlier times, when the world population of people was so much lower than in modern times.

Spraying a tank mix of fungicide and growth regulator.

Aerial Photography with a Drone. Start of Spring Seeding. 'Health & Harmony' Defra Consultation Paper

Crop Inspection and Recording with a Drone

At Kensham Farms we have a new toy, Charlie entered a competition promoted by one of the large fertiliser companies and won the first prize – a Drone, with quite a good lens, worth around £500. It is in the shape of a white cross, enclosing the battery and four small electric motors, measuring 400 mm from corner to corner, with a helicopter type blade 240 mm in length on each corner, so that the overall working width is 640 mm (just over 2 feet).

There is a handheld control panel fitted with a battery that has sufficient power for 20 minutes flying time, with a spare re-chargeable battery included and the whole apparatus, a DJI Phantom 3, is supplied in a protective carrying case with straps in the form of a rucksack. Alex has become most proficient at controlling it for crop inspection purposes, for which it has a range of 120 metres (400 ft) altitude and 450 metres (1/5th of a mile) from the controller. The photos show the Drone taking off in flight with Alex controlling it, and crop spraying in progress at Grove Farm.

Alex operating the drone.

Start of Spring Seeding

We have kept detailed rainfall records for the past 26 years which showed the 113.5 mm (nearly 4.5 inches) of rain in March to have been the wettest March during all those years. Normally we hope to have completed spring seeding of barley or wheat crops by the second week in March, but this year we were not able to start cultivations for this spring drilling of barley until 6th April, and completed drilling spring wheat on 20th April.

Similarly, our usual treatments for cereal crops seeded in the Autumn have been delayed by the wet soil conditions. This delay has not been quite so serious since those crops, mainly the winter wheat which we grow on around 80% of our fields, are growing well - but they do need spray treatments of fungicide to be applied at the correct growth stage to protect them from foliar diseases. This is important, since a diseased leaf cannot absorb sunlight properly, and if it fails to absorb this light it will not be able to grow properly through the process of photosynthesis. Photosynthesis from sunlight enables a healthy plant, with its leaves containing green chlorophyll pigments, to absorb water and carbon dioxide from the atmosphere and then convert them into carbohydrates for plant growth. As a by-product oxygen is released into the air.

'Health & Harmony' Defra Consultation Paper

Most farmers at the present time are worried about the future of farming and food production in Britain following the publication of 'Health and Harmony', the 64 page, but strangely titled, Government policy proposals for post-Brexit agriculture in Britain, put forward for consultation by the Secretary of State for Defra, the Rt Hon Michael Gove MP. We are taking this opportunity to comment on the proposals, since we feel that as they stand the proposals are fundamentally flawed by failing to mention food security for the nation.

The paper sets out five aims of the new British policy, which will supersede the EU Common Agricultural Policy (CAP), as being to provide public money for the following five 'public goods' described below: -

- *Environmental Enhancement and Protection*
- *Animal and Plant Health and Welfare*
- *Improved Productivity and Competitiveness*
- *Preserving Rural resilience, Traditional Farming and Landscapes in the Uplands*
- *Public Access to the Countryside*

Food Production is not listed in this 'Health and Harmony' paper as being a public good. So the implication is that the growing of food in the UK is not a public benefit, and that the security of food supplies coming from farms in Britain is of no importance, and that the production of farm produce such as milk, eggs, flour for making bread biscuits and cakes, wheat and oats for breakfast cereals, beef, lamb, pork, ham, bacon and barley for malting to brew beer from our own farms just does not matter.

Just over a year ago, in the Spring 2017 issue of Clarion V, the author described the objective of the Agriculture Act 1947, which set out its principles as having been: -

'To promote a healthy and efficient agriculture capable of producing that part of the nation's food which is required from home resources at the lowest price consistent with

the provision of adequate remuneration and decent living conditions for farmers and workers, with a reasonable return on capital invested'

That Act had been passed at a time when the danger of food shortages was a real thing, remembered by all those who had lived through World War II when they experienced the food rationing that had successfully eked out food supplies so that no one in this country starved. But now UK Government appears to be set on ignoring the lessons that were learned in those hard times and has the intention of overthrowing the 1947 Agriculture Act.

It is to be regretted that the author of 'Health & Harmony' appears never to have seen an empty shelf in a grocer's shop, and appears not to have a full knowledge of the workings of the global food market. Furthermore, the current position of UK agriculture shown in 'The Future of Farming and Environment Evidence Compendium' appears to have been ignored. This Evidence Compendium published in February 2018 by Defra and the Government Statistical Service is one of the most thorough reports on British farm economics the author has ever read.

The author's view, which seems to be shared by most of his friends who are readers of The Clarion, is that most farmed land in Britain does provide a pleasant landscape, and that it is often seen by those from the towns who visit the countryside, and that there is a good system of Public Footpaths (certainly in our Chiltern Hills area), and that folk can roam in all the remoter mountain and moorland areas which are now classified as 'Open Access Land', and that farmers treat their livestock with consideration (and indeed sometimes with affection), and that they take all the steps they can to protect their crops from plant diseases, and that there have been successful initiatives already introduced in recent years by the EU for environmental enhancement such as water purity in our rivers, air quality, protection of wildlife and maintaining land in good agricultural and environmental condition. It is a pity that the author of 'Health & Harmony' has not acknowledged these good things.

Drone photo of crop spraying in Sunters field, West Wycombe Estate, adjacent to the Adam Park Football Stadium.

The Clarion – *Autumn 2018*

Drought of Summer 2018

Summer Drought

The most common question that I have been asked in recent weeks is "how has the drought affected the farms", since the lack of rainfall this Summer has been astonishing - quite exceptionally dry. Our own records, over the 20 years from 1998 to 2017, show (in millimetres): -

	20 years average	2018
June rainfall	62 mm	4 mm
July rainfall	60 mm	16 mm

The July rainfall fell entirely during the last weekend of the month, so that we had nearly six weeks of the Summer growing season without any rain at all.

At Kensham Farms all of our large fields are used for arable cropping, with Winter Wheat grown for milling being our main crop, whereas our smaller paddocks close to the farm buildings are all used for DIY Horse Livery. If we look at the impact of an exceptionally dry summer on these enterprises under different headings we can summarize the different crop characteristics, and the effect of the summer drought on them, as being: -

Winter Wheat

There are 41 different varieties of wheat recommended by the UK Agriculture and Horticulture Development Board (AHDB) in the 2017/18 shown list that are suitable for seeding in the Autumn, normally in September or October. Of these varieties, only six are classified as 'Group I', meaning that they will be suitable for milling to provide the best kind of flour for bread making. At Kensham Farms we like to grow mainly these varieties of Group 1 milling wheat that are recommended by Warburtons as meeting its high standards for the bread which it bakes and sells through most shops and supermarkets. Warburtons buys all of its UK wheat through Openfield Agriculture Ltd, which is a large farmer co-operative based in Lincolnshire with an annual turnover of £700 million, and blend it with imported high protein Canadian red wheat. We have been members of Openfield, or one of the earlier farmer co-operatives which merged to form the large Openfield Co-operative since 1966. Currently we have entered into a five-year contract with Openfield to supply 2,400 tonnes of wheat each year for Warburtons, subject to it meeting strict quality standards.

Combine harvester cutting wheat, with straw chopper in action.

These quality standards include the protein level, ideally around 12.5% protein, and good 'bushel weight'. The 'bushel' was an old imperial measure of 8 gallons, used for grain and other animal feeds. A bushel of good quality wheat weighs more than the same measure of poor-quality wheat containing shriveled grains and husk. In modern metric terms the good sample of milling wheat should weigh not less than 76 kilograms per hectoliter, whereas the poor-quality sample, only suitable for animal feeds, might weigh 69 kg/hl. Additionally, the moisture content must be correct, not more than 15% moisture. In a wet season grain sometimes has to be harvested when it is as wet as 20% or even more, and then it has to be dried with warm dry air in the grain dryer. But this year we are not having that problem and expense - some of the grain is as low as 11% moisture content.

The colour of the crops at harvest time is also important, the ideal being a golden yellow colour shining in the sunlight of the harvest field. This year's sunshine has been good in that respect, with no trouble from the moulds or mycotoxins which can affect the straw and the seed head in a wet season. Furthermore, pre-sprouting of grain while still in the ear before harvest can reduce the important gluten content, and this sometimes happens in other years when it is wet in July or August. The gluten content of the grain can be tested with a clever test of the tenacity of the dough measuring the 'Hagberg Falling Number'. If the gluten content is low then the loaf of bread will not rise properly when it is baked, and in extreme cases might result in a stodgy loaf.

For our main winter wheat crop this year's summer drought has not reduced yields significantly below yields in an average year, since the seeds were sown last September and so the young cereal plants had established well before the winter. Then the crop was able to use the excess rainfall which fell in March, although it was difficult to apply the several necessary treatments of crop protection spray until April. Those treatments included herbicides for weed control, fungicides to prevent diseases like mildew, and growth regulator which makes the straw grow shorter but stronger, with a stronger root system. We were pleasantly surprised with our harvest results after the drought to find how little damage had been caused by the lack of summer rainfall, although on gravelly areas of the fields yields have been poor.

Spring Wheat

All of the varieties of winter wheat described above require a period of vernalization, that is a cold spell in the early stages of growth of the plant. This enables the plant to recognise that the winter has happened, and that the season for reproduction, by forming a new seed head, has started. However, many varieties of wheat have been developed which do not require a period of vernalization, and so can be seeded as spring wheat generally in late February or March. If winter wheat were to be planted in March it might not form ears of grain until the following year, since it would still be waiting for a spell of cold winter weather.

We only have a few fields of spring wheat and do not expect such good results as for the winter wheat, since more of their active growing season will have been during the period of drought.

Oilseed Rape and Other Crops

We also grow some fields of winter barley which matured early and have yielded well, and other fields of spring barley. The best quality barley is used for malting for brewing beer, with the lower quality samples being used for pig and other animal feeds. Oilseed rape is another crop for which there are both winter and spring varieties – this year we started harvesting the oilseed rape on Top Plain, the large field on the left of the B482 road to Stokenchurch, on 16th July – much earlier than most seasons. The small black oilseed rape seeds will be crushed to extract the oil for making margarine, mayonnaise, cooking oil and similar uses, as well as for biodiesel fuel.

Grass Paddocks

Our grass paddocks are used for DIY livery for horsekeepers – and the grass, just like everyone's lawn at home, and Hyde Park in London, stopped growing and turned brown during the drought, so the horsekeepers have had to supplement the grazing with extra fodder.

So in summary this summer for us, growing mainly winter cereal crops, has not been too different from normal summers. But for vegetable growers, and for dairy farmers and others with grazing livestock the drought has been exceedingly serious – and foods such as hay and silage saved for next winter have already had to be used to feed the animals during the summer.

The Clarion – *Winter 2018*

Ploughing Matches. Post-Brexit UK Agricultural Policy. Disastrous Agriculture Bill of 2018.

RSBAA Ploughing Match at the Berkshire College of Agriculture. The tractor in the foreground is a historic Fordson drawing a specially adapted match plough.

Ploughing Matches RSBAA – Encouragement of Good Agriculture and Livestock Husbandry

The Autumn is the time when most ploughing matches are held - in this area the Royal South Bucks Agricultural Association always holds its ploughing match on the first Wednesday of October. The South Bucks Agricultural Association was formed in 1833 *"to stimulate Farm Labourers and Servants to greater industry and skill in their several callings"*, and the following year King William IV graciously consented to become Patron of the Association and made a donation of £10 to its funds, thus conferring upon the Association the title of the *'Royal South Bucks Agricultural Association'*.

The RSBAA continues to encourage *'Good Agriculture and Livestock Husbandry'* on the farms within a ten miles radius of Beaconsfield, and awards to the winners of several different categories of crops and farm management classes the Association's silver cups as prizes to be held for the ensuing year. The author was elected last year as President of the Association for a two-year period of office, and in that capacity had enormous pleasure at the Association's annual lunch in October in presenting the Association's premier award for 2018, The Kings Cup for the best farm, to Daniel and Gideon Lacey of Laceys' Family Farm at Bolter End. The writer had equal pleasure in presenting to his own son, Charlie Edgley, the President's Challenge Cup for the best large farm in the Association's area for 2018.

The photos show the author presenting these awards at the Association's annual Ploughing Match and Lunch, attended by 400 members and their guests, at the Berkshire College of Agriculture at Burchetts Green.

Bryan Edgley presenting the President's Challenge Cup to his son Charlie for the Best Large Farm at the RSBAA Ploughing Match and lunch 2018, watched by RSBAA Secretary, Jo Short (on left), and by Alison (centre).

So for other residents in the catchment area of Clarion V, please be assured that we farmers in the Parish are doing our best to uphold good agricultural practice on our respective farms within the Parishes of Lane End and Cadmore End – and that there was no favouritism over the judging, since all the judges were farmers from other associations outside the South Bucks area.

Post-Brexit UK Agricultural Policy

The author regrets that he cannot continue this report with other optimistic matters that are favourable to the future of British agriculture. The Secretary of State for Defra, the Rt. Honourable Michael Gove MP, has recently published the Agriculture Bill 2018 – a Bill which the author considers to be a disgraceful attempt to downgrade the importance of the British farming industry.

In the Summer 2017 issue of Clarion V the author described the 1947 Agriculture Act passed by the then Agriculture Minister, Tom Williams, at the time just after the end of World War II when food in Britain was in such short supply that everyone had to be issued with a Ration Book. These Ration Books had to be presented at the grocer or butcher whenever the customer was buying the family's food to prove entitlement to that food, and that the family's food coupons for the week in question had not already been used. The objective of that 1947 Agriculture Act was to prevent similar food shortages ever happening again in Britain as a result of British farms being allowed to become run down or derelict.

The new Agriculture Bill will now make its way through Parliament, where we hope that amendments will be made to it, before it becomes enacted as the Agriculture Act 2019, since the production of food from British farms is hardly mentioned in the Bill. Around 70% of the landscape is cared for by farmers as a by-product of farming the land for the production of high quality food, and this is not likely to continue unless the Bill is modified in Parliament before it is enacted as the Agriculture Act 2019.

Since 1973, when Britain joined the European Economic Community, British Agriculture has followed the Common Agriculture Policy (CAP) under which all agricultural regulations emanated from Brussels, with UK government just having to carry out EU requirements. The result of this has been that no politicians in UK Government since 1973 have actually had to give thought to the strategic requirements of ensuring that the people of Britain would be able to buy sufficient food for their needs – but in future this ought to become the first priority for the Minister for Defra and the farmers that have to follow his regulations.

The sequence of events leading up to a new agricultural policy for this country has been: -

- February 2018 Defra publication of *'The Future Farming and Environment Evidence Compendium'* – a most comprehensive and accurate summary of farming statistics.

- February 2018 Defra publication of *'Health & Harmony: The Future for Food, Farming and the Environment in a Green Brexit'*- a disappointing policy document, which ignores the importance of food production from British farms.

- Defra then held a series of forty 'Farmer Engagement Meetings', one of which was held here at Kensham Farm at the end of February 2018. It was led by three competent young Defra officers, with 15 of us farmers present, to discuss the Defra policy outlined in its 'Health & Harmony' proposals. At that meeting we spent about four hours discussing the realities of farm economics, quoting statistics on growing cereal crops from our Kensham Farms' figures over a 23 year period, and an analysis of the 2017 figures from the 7,461 hectares (18,437 acres) farmed by members of the Chilterns Arable Group. We then formulated suggestions for future UK Government support to come into effect when the EU is no longer responsible for British farming.

- June 2018 Defra publication of a further competent report entitled *'Farmers' Voices, Government Listening'* – a good summary of the findings of the 40 'Farmer Engagement Meetings'.

- September 2018 introduction of the 2018 Agriculture Bill. This Bill has been widely criticised over its failure to attach any importance to the production of food from our farms, its failure to take note of global factors affecting the food chain, its failure to take into account the findings of Defra's own published statistics, and its failure to take note of its own summary of evidence from all the Farmer Engagement Meetings.

Disastrous Agriculture Bill of 2018

The writer considers this new Agriculture Bill to be a disaster which overthrows all the good aspects of the last UK Agriculture Act in 1947. It concentrates on environmental measures without recognition that farmers cannot 'be Green if they are in the Red'. The author hopes that Government will modify the 2018 Agriculture Bill so as to: -

- Encourage and support lowland farmers to produce food and care for the environment as a by-product
- Encourage and support farmers with Open Access Land *(such as mountains, moors and heathland where the general public have 'The Right to Roam')* to look after the landscape and environment and produce food as a by-product

The Clarion – *Spring 2019*

Food Shortages if 'No Deal Brexit'. Food Security. Spring Drilling.

Where does our food come from?
Origins of food consumed in the UK in 2017

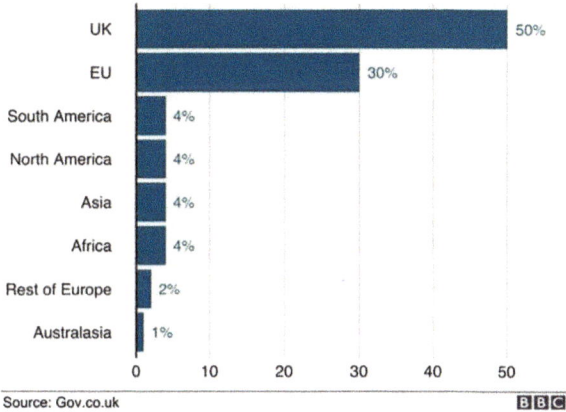

Region	Percentage
UK	50%
EU	30%
South America	4%
North America	4%
Asia	4%
Africa	4%
Rest of Europe	2%
Australasia	1%

Source: Gov.co.uk — BBC

In 2017 50% of the food consumed in the UK was home grown, and 30% from other member countries of the European Union.

Food Shortages if there is a 'No Deal Brexit'

Readers of Clarion V may well be reading this issue hoping to see local news, rather than national news, being exasperated with lack of decision and progress over Brexit. But it would be wrong not to mention the significance of the joint open letter that six of the major Supermarkets sent to the Government in January. In this open letter the senior executives of these major retailers of food make it clear that in the event of a 'no-deal' Brexit the result would be food shortages, empty shelves in the Supermarkets, and whatever food was available would cost more. This would be the inevitable sequel if the UK were to pull out of the EU without formulating a good policy for future trading with European growers and exporters, since 50 % of the food consumed in this country comes from our own British farms with a further 30 % coming from the EU (see the diagram from Gov.co.uk)

Food Security

Those of us who have been lobbying for a sensible future UK Agriculture Policy for post-Brexit UK farming will not have been surprised at this question mark over future security of food supplies in Britain for all those who live in the towns, as well as us in the rural areas. Many things in life are nice to have, but adequate food, wholesome drinking water and a place in which to live at a tolerable temperature are necessities for survival. My hope is that by the time this is printed our politicians will have reached an agreement with the European Union whereby trade with Europe can continue largely as at present, but with UK Government less constricted by EU legislation over other matters in the future.

Nigel Rogers loading seed into the seed drill in 1965, watched by Paul Edgley.

The new method in 2019. At seeding time Michael Leaver drives the Kramer forklift truck, drawing a trailer with 1 tonne bags of seed corn, which he then loads into the seed drill driven by Charlie.

The six essentials for the future of UK Farms are: -

- Avoid a 'No Deal' Brexit
- Ensure continuing free trade with Europe
- More discretion over regulations affecting farmers
- Maintain access for seasonal workers to come to UK
- Imported foods to comply with UK high quality standards
- Production of food to become key point of UK Agricultural Policy

Spring Drilling

Sowing the seeds for an arable crop with mechanical equipment is generally referred to as 'Drilling'. At Kensham Farms around 80% of our crops of winter wheat or winter barley are seeded in September or October, known as 'Autumn Drilling' with the other 20% this time of year as 'Spring Drilling'. The fields are normally too wet in the winter to take the weight of a tractor, so in most years Spring Drilling starts in early March when the fields are sufficiently dry following winter rainfall or snow. Last year (2018) there were 113 mm (nearly 4.5 inches) of rain in March, so no drilling could take place until April – and the danger of late drilling is that it can be followed by drought in the Summer. Our average annual rainfall over the last ten years has been 824 mm (32.5 inches), but monthly rainfall over that ten year period has varied from the lowest month of June 2018 with 4mm, up to the wettest month which was December 2013 with 140mm.

The very first seed drill drawn by a horse was perfected in 1701 by Jethro Tull who farmed in Berkshire. Before that date seeding was by hand from a trug basket or bucket – so it could be said that Jethro Tull's drill helped to bring about the British Agricultural Revolution. The design of seed drills has varied over the years – when we started farming Kensham Farm in 1955 all drills had hoppers for the seed which were the full width of the strip being seeded, often 6 or 8 feet wide. Many of those drills had two hoppers, the one for seeds and the other for fertilizer. The photo taken in 1965 shows Nigel Rogers (who was still driving our combine harvester during harvest 2018) loading one of the ICI fertilizer bags into the hopper of our Lundell drill of this design, watched by my son Paul then aged 7. The bags of seed can be seen at the front of the trailer – in those days the bags weighed 1 cwt each (ie 112 lbs or 50.8 kg) whereas in modern times the Government Health & Safety Executive (HSE) has set the maximum safe load to lift as being 25 kg for men, or 16 kg for a woman.

The second photo shows the modern design of seed drill, made by John Dale Drills in Lincolnshire, being loaded with seed corn in 1 tonne non-returnable sling bags which have a draw cord to open a spout at the bottom of the bag. The Kramer forklift is fitted with a trailer hook, so that its driver can load a trailer with six or seven bags of seed corn from our storage barn, haul them out to the field being drilled, and then unhook the trailer so as to load the drill, as shown in the photo.

These modern improvements of design and materials handling mean that in a working day we can drill the seeds on fields totalling around 140 acres (57 hectares).

Balance between Food Production and Care of Countryside. Control of Couch Grass. Work of Agricultural Research Scientists.

Balance between Food Production and Care of Countryside

My first *'On the Land'* jottings published in the March 2007 issue of The Clarion followed a Lane End Conservation Group column in the December 2006 issue in which the opinion was put forward that *'there was general agreement that modern farming and gardening practices played their part in upsetting the balance of nature, resulting in starlings and many other species getting scarcer'*. That author went onto state that *'If farmland is drenched with pesticides, then the food* (for birds) *is no longer present and the young die of starvation. Or, if the food is contaminated, then the young die of poisoning'*.

That article prompted me, greatly encouraged by the late Ross Osborne, to write something for The Clarion that described the work on modern farms, and the attempts by us farmers to keep the countryside where we live, and the wildlife in it, in as good a balance as is possible. This balance has to allow for the production of enough food to prevent starvation of the people throughout the world following modern huge increases of population, both in Britain and in overseas countries, and at the same time to look after bees, birds and other wildlife in the countryside.

Spraying of crop protection chemicals ('pesticides') is sometimes thought by the casual observer to be unnecessary, and yet it has resulted in average yields of wheat and other cereal crops having tripled in the years since I have been farming Kensham Farm. When Alison and I started farming here at Michaelmas 1955, just a few weeks after we married, at our first harvest in 1956 we were quite satisfied with any cereal crop that yielded over one tonne of grain per acre. Nowadays my son Charlie, and other farmers within the Chiltern Arable Group, are disappointed with any field of wheat or barley that is less than around 3.3 tonnes per acre.

Control of Couch Grass

It was immensely difficult to control couch grass in cereal crops before the development of effective herbicides. At that time couch grass, as any gardener knows, is a most troublesome weed grass that spreads primarily through the rhizomes of its underground root system, so that the root itself puts up new plants. The way of eradicating couch grass before the advent of suitable sprays was to cultivate the field so that the rhizomes of couch came up to the top, followed by harrowing the field with chain harrows to roll all the rhizomes up into bunches. Hand work was then necessary with a 4 tined long handle fork to pick up all the bunches of rhizomes and make a series of small bonfires over the whole field. Nowadays, if there is any couch grass in a field it can be sprayed with Glyphosate between crops so that the active ingredient is translocated from the leaf down to the root thus killing the entire plant and making the ground suitable for the next crop.

Charlie checking one of the outer nozzles of the Amazone Pantera self-propelled sprayer.

As a point of interest, I have looked up my copy of Fream's Elements of Agriculture which was the standard text book on farming when I was at Agricultural College in 1954. This textbook, running to 715 pages, was revised in 1951. The reference to weed control by chemical methods gave the advice on the use of Sulphuric Acid to control weeds, a practice which had started in the 1920s. This acid was referred to as BOV (Brown Oil of Vitriol) and was highly corrosive and dangerous to handle. Fream's went on to make a short reference to research started in 1941 on 'Growth Promoting Substances', sometimes referred to as 'hormones' such as MCPA, which worked by making the weed outgrow its own strength before dying.

Work of Agricultural Research Scientists

The dramatic increase in production in the last 40 years has been partly due to the breeding of higher yielding varieties, but the greatest increases have come from artificial fertilisers and the use of crop protection chemicals developed by research scientists. These crop treatments, which are all applied with great precision to protect the health of the crop, include : -

• **Seed Dressing**. Cereal seeds are always treated by the supplier of the seed with a coating of seed dressing before delivery to the farm. The purpose of the seed dressing is to protect the very young plant as it emerges from the soil from various fungal diseases. Some seed dressings also contained neonicotinoid insecticide which protects the emerging crop from attack by aphids, thus guarding the crop against the disease Barley Yellow Dwarf Virus which can cause the crop to fail. Most unfortunately the use of neonicotinoid insecticide for seed dressing became illegal in December 2018, so that in future it will only be possible to control aphids with greater quantities of different spray treatments after crop emergence.

- **Herbicides** may be selective weed killers, which can kill yield suppressing weeds in the crop without harming the crop itself. Some herbicides are applied by spraying the young crop and the weeds in it – these are called 'post-emergence herbicides'. But scientists have also developed 'pre-emergence herbicides' which are used to spray the crust of the soil before the crop has emerged – then the young cereal plant will grow through that crust of treated soil, but the harmful weeds will be killed as they emerge. Total herbicides such as Glyphosate are most useful for cleaning fields between crops.

- **Fungicides** protect the growing plant from fungal diseases such as Mildew or Septoria, in the same way that roses in the garden can be protected from Black Spot with a suitable fungicide. There are several different fungicides suitable for successive growth stages.

- **Growth Regulator** sprayed on the leaf of the plant at a midway stage of its growth will strengthen the root system and will make the stem of the cereal plant grow shorter but stronger, so that in a wet season the crop will be less likely to lodge (fall over). Since development of these growth regulators it is now unusual to see cereal crops which have been laid flat before harvest as a result of outgrowing their own strength.

- **Insecticides** are never used as routine, but only for specific infestation of a crop by aphids, flea beetle and similar pests.

- **Trace Elements** the main plant nutrients of Nitrogen, Phosphate, Potash, and correction of soil acidity with Lime, are all spread as solids to the fields. However, crops can sometimes be short of trace elements such as Manganese and these are best applied with the crop sprayer.

This summary would be incomplete without mentioning that some farms are managed on 'organic' principles under which the only crop protection chemicals used must be within the limited range of such products that have been approved for use by the Soil Association. The present Minister for Defra makes generous subsidies available for organic farmers, but those subsidies are necessary to cover the greater costs and significantly reduced yields of organically grown crops. Many folk regard organic food as being top quality since it is more expensive, and yet no evidence has been shown to prove any additional benefit to human health from foods grown on organic principles.

The Amazone sprayer, with boom width of 24 metres (80ft) spraying a tank mix of fungicide and growth regulator.

Big Farmland Bird Count

On a foggy morning in February of this year, we were pleased to welcome an ornithologist from Berks, Bucks & Oxon Wildlife Trust (BBOWT) to assist us with surveying for the 'Big Farmland Bird Count', an initiative of the Game & Wildlife Conservation Trust. We were pleased to record several Yellowhammer and Linnet as well as Skylarks. Across 9 farms in the Central Chilterns Farmer Cluster there were 297 recordings of 59 species including the 'champagne moment' of spotting a pair of Short Eared Owls near Lacey Green.

The Clarion – *Autumn 2019*

Harvest 2019. Farming Calendar. Brexit and future of British Farming. RSA Food and Countryside Commission.

Harvest

We started harvest on 22nd July this year with our crops of winter barley – this is always the first crop to ripen. We have been pleased with the yields which have averaged 3.49 tonnes of barley per acre, with our best field yielding 4 tonnes per acre. Since the advent of 'hybrid' barley varieties approximately 6 years ago, yields have increased tremendously – in the early 1960s we would have been pleased with yields exceeding 1.5 tonnes per acre.

The photos taken with our drone were at Kernals Field on the West Wycombe Estate, at the side of the A40 road between West Wycombe and Piddington. Growing conditions this season have been satisfactory, with a good start in October 2018 when these crops were seeded into a firm dry seedbed. In a wet Autumn, when the fields are sticky from rain which has fallen but not dried off, the crop is generally off to a bad start since the tractor and seed drill wheels are apt to press down on the wet soil, leaving a crust which inhibits growth of the young emerging plants.

A typical summary of the work that goes into an Autumn sown crop of wheat or barley is: -

September – Spreading farmyard manure (FYM) or treated sewage sludge on the stubble will be the first job for the new crop. Since May 2018 we have had a good arrangement with Laceys' Family Farm whereby we supply Laceys with straw in swathe from our cereal crops for them to bale and use for their dairy herd. In exchange Laceys supply the FYM that is surplus to their own requirements for use on our arable fields. The stubble from the previous crop is then cultivated to encourage germination of any weed seeds, before killing out these young weeds with Glyphosate.

October - Sowing the seed with an 8 metre wide seed drill. This seed will have been treated with a seed dressing to protect the young plant as it emerges from the soil from various fungal diseases.

Neonicotinoid insecticide used to be included in the seed dressing, to protect the emerging crop from attack by aphids. Regrettably the use of neonicotinoids has now been made illegal by the EU, so Autumn 2019 will be the first season without this protection. This means that Autumn spraying with insecticide will be necessary if the growing crop should suffer from attack by aphids.

The seed bed will then be sprayed after seeding with pre-emergence selective herbicide.

This will kill young weed seedlings, without harming the cereal seedlings, as they poke through the soil crust. The field must then be watched for signs of damage by slugs, to be treated by application of slug pellets

March – Spreading nitrogen + sulphur fertilizer. (Potash is normally only spread in alternate years, whereas the Phosphate necessary for the crop is generally provided by a dressing of sewage sludge once every 4 or 5 years)

April - The crop is likely to be sprayed firstly with a mixture of Growth Regulator and Fungicide, then later in the month with a further spray treatment of Growth Regulator mixed with a different type of fungicide.

May - A spray treatment of herbicide with a third type of fungicide. These fungicides are necessary to keep the leaf in healthy condition, without damage from fungal diseases such as Mildew or Septoria. A healthy plant leaf can then use the summer sunlight for photosynthesis, the process whereby carbon dioxide from the air plus sunlight on the leaf forms the glucose from which the growing plant builds itself and the following year's seeds within it.

June – A further treatment of fungicide, designed for use late in the season.

Late July or August – Harvest time.

Drone photos of harvest at Myze Farm on the West Wycombe Estate. It can be seen that the grain tank on the combine is almost full, with the grain trailer about to return to be filled.

Brexit and the Future of British Farming

National news on television and in the papers has been dominated this Summer and Autumn by Brexit, with the election of Boris Johnson as Prime Minister during July. For farmers the important appointment is that of the new Minister for Defra, the Department for Environment, Food and Rural Affairs who will be Theresa Villiers, assisted by George Eustice who will be returning to Defra as Farming Minister.

The importance of these new Ministers in Defra, as well as the attitude of the new Prime Minister towards farming and the rural areas, will be more crucial to us farmers than in the past. The reason for this is that while we were within the European Union farm policy was formulated in Brussels, so that the English Defra Minister was compelled to follow those overall EU policies. If we do leave the EU, which is highly probable but not a total certainty when this issue of Clarion V goes to press, it will be the first time since 1972 that British farming policy will be set out in Westminster rather than in Brussels. Farmers continue to be worried that there could be an influx of low-quality imported food in the event of a 'No Deal' Brexit.

The RSA *Food Farming and Countryside Commission*

The RSA, an abbreviation for the Royal Society for the Encouragement of Arts, Manufactures and Commerce was founded in 1754 – so one could say that it is now well established. Its mission is to address today's most pressing social challenges – such as the economy, employment, education, health service and prisons.

The RSA transferred its interests in Agriculture to the Royal Agricultural Society of England in 1840, but in November 2017 the RSA set up its *Food, Farming and Countryside Commission* to look at both the farming industry itself, and at the ecosystems within which it operates, and climate change, and matters such as diet related ill-health.

In October 2018 we hosted a meeting with one of the RSA's bike tour reporters, this was part of a nationwide tour to meet farmers and rural food businesses using pedal power. Some findings of this RSA report are: -

- 72% of UK land is farmed, but only 1% of the UK workforce is employed in agriculture
- By 2030 there will be 9 billion people in the world, many of them in India and China, all of whom will need food
- Healthy foods must be at the heart of the future of the UK food system
- The production of good healthy food should become good business
- There should be a 10-year transition to 'agroecology' with measures such as planting trees and restoring natural grassland, with less reliance on modern crop protection chemical treatments

Some of us who have the day-to-day task of producing healthy food from our farms will watch with interest to see whether future regulation from Defra follows up these RSA recommendations, in the context of the world's increasing population and global food system.

Discharging the spring barley in the tank of the combine into the 16 tonne capacity trailer, on the run without stopping.

The Clarion – *Winter 2019*

Crop Rotation. UK Agriculture since World War II. Autumn Seeding. Brexit.

Autumn Arable Work

Since the last issue of Clarion V we completed the 2019 Harvest on 29th August following a good sunny August apart from a short spell of rain in the middle of the month. Our yields of grain were good this year, with the heavy rain in early June sufficient for the cereal plants' needs.

Our main crop is winter wheat with 555 hectares (1,363 acres) yielding 5,227 tonnes – most of it being milling wheat suitable for milling into flour needed for making bread. This yield of 9.42 tonnes per hectare (3.8t/acre) was one of our best years ever. When we started farming Kensham Farm at Michaelmas 1955 such a yield would have been considered impossible – at that time 3.7 tonnes per hectare (1.5t/acre) would have been considered as excellent.

September is the month between harvest and seeding the next year's autumn sown crops when jobs such as muck spreading can be carried out on stubbles, and a few of the fields are ploughed, some are cultivated, and others are just sprayed with Glyphosate to kill out any weed seedlings or young cereal plants. These can grow from light grains (known as volunteer seedlings) which were shed out of the back of the combine harvester. This is all in preparation of the fields for seeding the next year's crop in October.

Crop Rotation

I am sometimes asked what crop rotation we follow on our arable fields – the answer to which is that on many of our fields we grow continuous winter wheat, year after year, without any rotation of crops.

If we take a look right back into history, we find that in earlier centuries the problem was always how to grow enough food to feed the increasing world population. In medieval times peasants grew strips of crops, often in a three-year rotation. Then much land was enclosed by fences or newly planted hedges, so by the mid-1800s most good productive land had been enclosed.

There had been a great improvement in production of food from the land in the early 1700s when Charles Townsend developed a rotational system of cropping known as the 'Norfolk Four-Course System'. This was a sequence each four years of *wheat, turnips, barley and clover*. Each crop in this 4-year rotation had a purpose: -

- The wheat was grown for milling (at those former times by windmills, or by the force of a stream turning a water wheel connected to millstones) to make bread.
- Following the wheat, a root crop was grown such as turnips or swedes or mangold-

wurzels as feed for the cattle for the next winter. Since the root crop would have been hoed by hand or with a horse hoe, this controlled the weeds which would have seeded in the preceding year's wheat crop.

- After turnips or other root crop, the next year barley would have been grown. This was used to feed pigs which were reared for ham and smoked bacon, and also for malting to brew ale and beer.

- After the barley a clover ley was grown for grazing by sheep in fields or by cattle. This also controlled the weeds that would have multiplied in the barley crop, and since clover has root nodules which fix atmospheric nitrogen to form nitrates, these nitrates would have fertilised the soil ready for the wheat crop at the next 4 season sequence of cropping.

UK Agriculture since World War II

The earlier Four-Course sequence of cropping was often used on mixed farms right up to the end of World War II. Then in the post war years research scientists have developed effective herbicides to kill weeds, and fungicides to prevent leaf diseases such as septoria or mildew and have improved the manufacture of inorganic fertilisers. So nowadays the herbicides effectively kill any weeds, without workers having to remove them with a hoe in the root crop. Fertilisers now provide the nitrogen that in earlier years would have been fixed from atmospheric nitrogen by the root nodules of the clover crop.

The result of these developments is that farms nowadays specialise more. Many arable farms on the Chiltern Hills and towards the Eastern Counties now crop the land with cereal crops on a continuing year by year basis. Farms towards the West of England and in Wales, where average rainfall is higher, often concentrate on rearing cattle for dairy produce or beef and sheep for the production of fat lambs, with wool as a by-product. That is why most of the fields in the West are pastures of permanent grass, often with wild white clover in the sward.

Autumn Seeding

The photos show our two seed drills at work during October, which has been a difficult month with considerably higher than average rainfall. The smaller Weaving drill at work on Myze Farm, with West Wycombe House in the background, is mounted on the hydraulic linkage of the 215-horsepower tractor. It seeds a width of 6 metres, the hopper holds 1 tonne of seed, and it has a spot rate of around 12 acres per hour.

Much of the weight of the larger Dale drill, shown here at work at Fillingdon Farm, is carried by its own wheels, with some of its weight transferred to the back wheels of the heavy 360 horsepower tractor which is necessary to draw it. The seeding width of this larger drill is 10m (32.8ft), the hopper holds 5 tonnes of seed and has a spot rate of around 22 acres per hour. This has enabled us to make the best of the very few dry days this autumn, with 230 acres seeded on the best day at the end of October.

Brexit

Most of us are tired of news about Brexit and the impending General Election – but those of us who work in the farming industry all hope that the next Government, whatever its colour, will recognise the importance of good quality home-grown food.

Weaving mounted seed drill, with which the whole weight of the drill is carried by the tractor, West Wycombe Park in the background.

Horsche trailed seed drill seeding winter wheat.

The Clarion – *Spring 2020*

Agriculture Bill 2020.

The new Agriculture Bill has been described in the Farm Business weekly bulletin as being *"The biggest shake-up in domestic farm policy in living memory"*.

Farmers meeting with Steve Baker MP at Kensham Farm, Charlie on his right, Alison and Bryan on his left.

Agriculture Bill 2020

This new Bill was announced in January 2020 and has been presented to the House of Commons for scrutiny and debate by all Members of Parliament. When it has been agreed with any amendments in the House of Commons it will then be passed to the House of Lords for further discussion and comment before being enacted as law. At that stage it will become the Agriculture Act 2020 - the first such new legislation since the Agriculture Act 1947.

In an earlier edition of Clarion V the merit of the Agriculture Act 1947 was described. It had been introduced by the then Labour Minister of Agriculture, Tom Williams with these words: -

"to promote a healthy and efficient agriculture capable of producing that part of the nation's food which is required from home sources at the lowest price consistent with the provision of adequate remuneration and decent living conditions for farmers and workers, with a reasonable return on capital invested".

That was just after the end of the severe shortage of food during World War II which the writer remembers well, with food rationing and Government books of Food Coupons which were issued to everyone in the UK from the beginning of the war until 1954.

Those food shortages were fresh in everyone's minds at that time. They had been caused by German U-Boats sinking the convoys of ships carrying food from overseas to the UK - the second time that it had happened, since enemy U-Boats had been operating in World War I and sinking so many ships importing food especially in 1916 and 1917, that we almost lost the first world war as a result of starvation of the folk at home. Then the same thing happened again during the early 1940's - since between the two wars Government had allowed the UK Agricultural Industry to become so run down that some farms became almost derelict.

Those were the reasons why the Agriculture Act 1947 was introduced, but now it is effectively being repealed by the Agriculture Act 2020 – with its new policy for farming, drafted by our politicians who are too young to remember food rationing, and who lack the caution to realise that the same thing could happen again.

This new policy for the future of farming in England, which is set out in this Agriculture Bill, appears to be based on the false thinking that we 'would always be able to import plenty of food from overseas in the future, and that Global Warming and the Environment are the only important things with which Defra, the Department for the Environment, Food and Rural Affairs, should concern itself.

Global Warming is indeed happening and will be a concern for us all in the foreseeable future. Politicians, and the doom-mongers at the recent Intergovernmental Panel on Climate Change jamboree at Davos, seem certain that the cause of global warming is carbon produced by mankind. But many scientists do not share that certainty, some of them point out that the energy output of the sun is not constant but varies over a time scale of tens of thousands of years causing an impact on our climate, and that the warming of recent years may precede a cooling phase. Rocks record evidence for past climates, including extreme conditions that have been linked to mass extinctions.

One certain factor is that there are now an unsustainable number of us humans living on the planet, and that we are using the earth's natural resources, such as coal, oil, metals and minerals, and in the agricultural industry phosphate for fertiliser, at a rate that is bound to lead to eventual exhaustion of those natural resources. It would be very wrong for politicians to suggest thinning out the number of humans on the planet – but time will tell if nature finds its own way of achieving that ultimate way of making life on earth sustainable.

One primary task for politicians should be to ensure a reliable supply of high-quality food for their electors. This is the same task that we farmers want to achieve. But it will be hard for British farmers to compete in an unregulated market with imports of food from other countries where governments subsidise food production but have less regulation and lower standards.

The new Agriculture Bill has attached more importance to the growing of food than was shown in Michael Gove's initial 2018 Consultation Paper on the future of UK agriculture, so it has been good that farmers' lobbying through their Members of Parliament (the photo shows those of us at Kensham Farm with Steve Baker MP after one such meeting in May 2019) has instigated some good changes within the future policy now set out in the Agriculture Bill.

But there are still most serious shortcomings of the new Agriculture Bill. Some of the main points within the 96 pages of the Bill, are listed below under the headings of those aspects farmers consider to be good or bad compared to the earlier EU farm policy are:

Good Points are the intention for: -

- Government support to become less bureaucratic than in the EU

- The contribution that farmers make to the environment to be recognised, with the introduction of a new Environmental Land Management Scheme, abbreviated to 'ELMS'

- Government to become more collaborative in developing a new support system

- Changes to be brought in gradually, phased over a seven-year period

- More opportunities for new entrants into farming as a career

- Policy, including food security, to be reviewed at 5 yearly intervals

Unfortunate Points are: -

- The food production side of farming is given insufficient backing – Government will only consider any support in the event of *'severe disturbance in agricultural markets'*

- Trade policy may allow imports which are cheaper, but have been produced using methods not permitted here

- Most countries other than New Zealand support farmers, so that effectively food can be provided in the shops for consumers at less than cost of production. Direct farm support in England will cease under this new Agriculture Bill, so British farmers are likely to face unfair competition

- Defra statistics *(The Future Farming and Environment Evidence Compendium)* published in 2018 show that most farms depend on direct support payments in order to stay in business. When this is cut off, farmers will have to cut corners to balance their books, and the first corner to be cut is likely to be farmers' current work maintaining the environment and landscape of the rural areas.

The photo shows the May 2016 Chiltern Vintage Tractor Run crossing Kensham Farm, with the Public Footpath STC/52/3 crossing in the valley between Pound Wood and Leygroves Wood.

Chiltern Vintage Tractor Run crossing Kensham Farm on its annual run of around 20 miles raising funds for the Thames Valley Air Ambulance.

The Clarion – *Summer 2020*

Covid-19 Pandemic. SARS 2003. Reappraisal of Principles of Globalisation. Spring Seeding 2020.

The Importance of Food Production.

In the last issue of Clarion V we were all considering the impact of Brexit, and for farming under the new Agriculture Bill 2020 which was intended to show the post-Brexit way forward for British farming in the years to come, with a greater emphasis on environment and conservation than on food production.

Many of us in the farming industry have felt that in recent years Government had lost sight of the importance of food, feeling that the world system of globalisation would always provide plenty of food, and that British farms should concentrate on caring for the environment as a primary objective, while the food we needed could be imported.

Covid-10 Pandemic

But we now have the situation which Bill Gates, founder of Microsoft, forecast in 2015 in a lecture convened by TED, the USA not-for-profit organisation founded in 1984 to arrange major conferences in the three fields of Technology, Entertainment and Design. In that lecture, five years ago, Bill Gates forecast that *'if anything kills over 10 million people in the next few decades, it's most likely to be a highly infectious virus rather than a war'*. He went on to state that *'we need lots of advanced R&D in areas of vaccines and diagnostics'*.

SARS 2003

Governments worldwide have not encouraged or financed that research into development of suitable vaccines or methods of testing for virus infections or the eventualities forecast by Bill Gates. If such research had taken place, which could have started after the outbreak of Severe Acute Respiratory Syndrome ('Sars') in 2003, then suitable vaccines for Sars would be very similar to the vaccine that is now needed to give protection against the threat of Covid-19.

We have seen what empty Supermarket shelves look like as a result of normal trading procedures being upset as a result of the spread of Coronavirus, which had started in a 'wet market' in the Chinese town of Wuhan from bats or wild animals being sold for human consumption. This virus has now spread worldwide, with no vaccine yet developed and ready for use.

Charlie, wearing suitable gloves and eye protection, loading a mix of crop protection chemicals, fungicide and growth regulator, into the loading hopper of the Amazone sprayer.

Reappraisal of Principles of Globalisation

We farmers hope that there will be a reappraisal of the principles of globalisation, and that the food which we produce will in future be valued in the way encouraged by the Prince of Wales in a recent article in Country Life in which he stated that '*Food does not happen by magic. If the past few weeks have proved anything, it is that we cannot take it for granted'*. We hope that while the Agriculture Bill is still under discussion in Parliament it will be amended further to reflect the importance of a reliable supply of home-grown food. We hope that support for food grown in Britain will not be another thing like the development of vaccines that Government puts off for another day, by which time it might be too late.

At the end of April we were still in the lockdown situation, but the work of growing food goes on, and the crops to be harvested in late Summer 2020 will be as essential as ever. The main precautions that we have had to take have been over ordering essential crop protection chemicals, many of which are manufactured in Germany and other countries including China. In a normal season we order only a few days in advance, so that we can be certain which of the alternative products will be the best to use for the prevailing weather and stage of crop growth. But this year we have had to guess in advance the most probable products to be required, and then place a bulk order in the hope that it will be delivered.

On the farm machinery side, the diesel fuel on which the tractors run is cheaper than previously due to lack of demand from cars and commercial lorry traffic. However, this has an unexpected downside for us in that the demand for US Maize Ethanol has fallen from 7.5 million barrels per week to only 4.0 million barrels. This has greatly lowered the demand for the raw ingredient of maize, with consequent fall in its market price, and the price of the wheat for breadmaking which we sell in the UK is linked to this world trading price of maize.

Spring Seeding 2020

Closer to home, the farm has suffered from the same most unusual extremes of weather that have affected gardeners. We had an exceptionally wet winter, with above average rainfall in every month from October 2019 to February 2020, followed by very little rain in March and most of April. The result of the wet Autumn was that we only completed around 60% of the planned Autumn seedings, leaving more fields to be seeded in the Spring – work that started on 12th March. We were thankful that 20 mm of rain fell in the night of Friday 20th April, just enough to keep the spring seedings of wheat growing nicely.

The photos show Spring drilling underway on the West Wycombe Estate land at Grove Farm and Bullocks Farm, for which we used the new technique of 'direct drilling', that is drilling into soil that had minimum disturbance or cultivation since the previous crop. The tines on the drill are a 'wearing part' of the drill – they wear out fast on our flinty soil on the Chiltern Hills, so in these uncertain times we must order and carry a good stock of replacement tines. Many metal wearing parts for farm machinery have to be imported from overseas countries such as China where they are forged.

Direct drilling Spring wheat in March 2020. After harvest 2019 the field was lightly cultivated with our Horsch Joker to encourage growth of any weed seedlings, prior to spraying with the total weedkiller Glyphosate (Roundup). No other cultivation was carried out prior to seeding, shown in the photo.

The Clarion – *Autumn 2020*

Cereal Crops. The Author's Military National Service. Bryan and Alison's early years at Kensham Farm.

Crop Production at Kensham Farms

Coronavirus and lockdown or not, the farming year continues – and with it the need for the crops which we produce, which this year will be: -

(i) **Wheat** for milling to make flour for bread and biscuit manufacture – this will be the preferred use for the best quality samples, with high protein and with the high gluten content so that the loaf of bread will rise nicely (whereas if the gluten content were to be too low any bread made from it would be stodgy and unappetising). Any poorer samples of wheat not coming up to these high standards will be used for the manufacture of poultry and animals feeding stuffs.

(ii) **Barley** – the best quality samples will be bright with low protein suitable to make malt, from which beer is brewed and whisky can be distilled. So it is interesting that on the farms we have to use amounts of nitrogen fertiliser likely to lead to low protein in barley, whereas for wheat it is just the opposite with high protein in the sample being the objective for top quality.

(iii) **Oats** – we sometimes grow oats but not this year. There is not a big market for oats. The best quality samples are used for making muesli, porridge oats and oat cakes, whereas low quality oats are used mainly for cattle and horse feeds.

The photo shows the start of our barley harvest this year – it was taken on 23rd July 2020 when we were harvesting barley off the sloping section of the field at Bigmore Farm known as Niddles. The combine harvester is a John Deere Hillside model, the photo shows how the cab with all its sieves and thrashing mechanism stays upright while the cutter bar adjusts itself to the contour of the field.

At the time of writing, we have had to pause harvesting for a few days following the 22.5mm of rain (nearly 1 inch) which fell on Sunday 26th July – exactly the wrong time for heavy rain, so soon after the start of harvest when the crops needed sunshine rather than rain.

Harvesting winter barley in July 2020.

The Author's Military National Service

In the last issue of Clarion V, Katy Dunn suggested that those who were finding that their jobs were not leading to fulfilling lives could spend some of their time and thoughts during lockdown on considering whether they could improve their happiness and life by making cautious and well considered plans to change direction.

This comment made me wonder whether my own past decisions in life might be of interest, or even of help, to some readers – since in the 70 years between age 18 and age 88 I have made one compulsory job change, then a job change of my own choosing, and then various changes of emphasis within that chosen career.

Firstly, all young men who were passed as medically fit in 1950 had to serve their National Service of two years in the armed forces. In my case I had always been interested in mechanical things and in rifle shooting, so I chose to serve in an Armoured Car regiment, the Royal Horse Guards which, although it still had a squadron of cavalry horses stabled in barracks in Knightsbridge for ceremonial duties of guarding HM King George VI in London, its real modern work was as an armoured car reconnaissance regiment trained for service overseas in times of conflict. After training at Windsor and Aldershot and being commissioned at Mons Officer Cadet School at Aldershot, I was posted out to the regiment which was stationed with the British Army of the Rhine (BAOR) in Germany at Wolfenbüttel near Brunswick. Our brief was to carry out border patrols along the "Iron Curtain" between West and Soviet occupied East Germany. These patrols were carried out by a troop of armoured cars consisting of a 4 tonne Dingo (a relatively small open topped armoured car) with crew of two, followed by

The author's troop of Daimler armoured cars in the Royal Horse Guards on manoeuvres in Germany.

myself in a Daimler armoured car with gun turret, then a civil servant from the Control Commission for Germany (CCG) in a black Opel car with a flag flying on its bonnet, then a second 7 tonne Daimler armoured car for my troop Corporal of Horse (the Household Cavalry title for a Sergeant) with his driver and radio operator/gunner and then a second Dingo.

It was while we were serving in Germany that we heard the sad news that the King had died. My two years National Service ended in October 1952, so this was my first career change – to civilian life. I then followed my father and grandfather into the legal profession by working as a solicitor's articled clerk in the City of London for just under a year – but it did not take many months for me to realise that life as a London solicitor for me had few attractions.

So then I had to decide what to do, whether to carry on or to make a bold change. After exploring the possibility of serving in the Colonial Police Force in the part of Africa that was then known as Tanganyika (modern day Tanzania), I decided on a career in farm management and learned the trade firstly by working for a year on a very well managed farm near Peterborough owned by William Abbott – from whom I learned how to think as a farmer. I also learned the technicalities of farming at that time, a time that now belongs to the history books, with seven working horses and self-binders for harvest.

Bryan and Alison's Early Years at Kensham Farm

Towards the end of that year, 1953 I met my late dear Alison on a blind date and asked her to marry me four days later, and to my delight she agreed. We announced our official engagement a few months later while I was at the Royal Agricultural College (RAC) at Cirencester, we married in August 1955 and then moved to Kensham Farm with its 102 acres and near derelict farmhouse in September 1955 on Michaelmas Day.

In those early years we lived in the same farm cottage that my grandson Alex now lives in. The farmhouse was subject to two official orders; firstly it was listed as a building of historical interest, and secondly it had been condemned by the local authority with a Slum Clearance Order as being not fit for human habitation. The effect of these two Orders was that we were not allowed to knock down the farmhouse, but we were not allowed to live in it until it had been renovated with a plumbing system and indoor WC lavatory. That work took around a year and a half.

During the early years, our dairy herd of Friesian cows for milk production for the Milk Marketing Board was the main enterprise, coupled with hatching egg production on contract to a newly established branch hatchery for day-old chicks near Wargrave. This was a branch hatchery of a Yorkshire firm, Thornbers.

In 1960, we were able to double the acreage by buying the adjoining Watercroft Farm, and in March 1970 we took on a Farm Tenancy of Dells and Bigmore Farms, leading to a switch from milk production to cereal cropping on the four amalgamated farms which by then totalled around 405 acres.

That was soon after the time when I was moved to change emphasis in my own life, with other interests as well as farming. Alison and I found great satisfaction in the work of and worship at, Holy Trinity Church in Lane End where I was first elected as Churchwarden in 1972, and I became a part-time Tutor at HM Borstal Finnamore Wood teaching maths, photography, group discussion and tractor driving – work which I continued for 22 years.

Meanwhile. Charlie joined our farm partnership in 1985 following a 3-year course at the RAC Cirencester, and under his initiative we started farming the major part of the West Wycombe Estate from September 1999 under a Farm Business Tenancy from Sir Edward Dashwood. We also took on various smaller Farm Business Tenancies (FBTs) and Contract Farming arrangements on other neighbouring farms, so that now we farm around 2,500 acres (1,011 hectares) in total, of which 2,000 acres (809 hectares) are arable cereal crops.

Bryan and Alison photographed on their last holiday together on the Isle of Wight.

It has been hard to be without my Alison who died in June this year, since we had done everything together, both for our family and the farm, for all these years. We had made plans for a family tea party in August to celebrate our 65th wedding anniversary, but that was not to be. The photo was taken on our last holiday together to the Isle of Wight in September 2019, a few weeks after our 64th wedding anniversary.

The Clarion – *Winter 2020*

What happens after harvest? Direct Drill with Tined Cultivators. Kensham Farms Foreman. Airfield for Model Aircraft.

What happens after harvest?

I am sometimes asked 'what happens after harvest' – as though that is the time when arable work eases off. But that is far from the truth, since with modern arable farming practices the year no longer has the steady sequence of one specific task for each season. Nowadays for the best yields of wheat, the staple food of the western world from which bread and biscuits are made, many of the tasks of growing fall within a three-month period, from the end of July to the end of October.

Harvest this year started on 21st July and ended on 26th August – it was not such a good harvest as in 2019, since there had been excessive rainfall in Autumn 2019 that hampered Autumn drilling ('drilling' is really another word for seeding). Because the fields were so wet at that time, we were only able to seed 60% of the planned number of fields. An ideal seedbed for drilling has a moisture content such that you could walk over the field wearing shoes. If the field were to be so moist that gumboots were essential, with wet soil caking on them, then the tractor would cause too much compaction, the seed drill would often clog up, and if the seeds were nevertheless to be planted then germination and subsequent growth would almost certainly be disappointing.

The remaining 40% of the fields that should have been seeded in Autumn 2019 had to wait until Spring 2020, and Spring plantings seldom do as well as Autumn seedings. The result of this was that yield of grain from the harvest 2020 crop was 2,000 tonnes less than the previous harvest – only 5,000 tonnes grown instead of the 7,000 tonnes which we grew for harvest 2019. However, I am pleased to report that this Autumn we completed the Autumn work of seedbed preparation and drilling the seeds by 20th October in good conditions – so hopefully this will be the foundation of a good harvest in 2021.

Direct Drill with Tine Cultivators

Our new 10 metre wide Dale direct drill, with tines rather than discs to make the narrow channel, about 2 inches deep, into which the wheat seeds are placed performed well and did not cause the type of blockages which had often hampered the work with the disc type drill that we had used in earlier seasons.

Another job which has to be carried out after harvest but before Autumn drilling is to cut those hedges which bound the fields to be cropped. For this we have bought a new Bomford Hawk hedge trimmer with flail action cutter driven by a hydraulic motor from the tractor's hydraulics, shown in the photo being operated by my grandson Angus, who has taken over that particular task from Nigel Rogers.

Angus trimming hedges on the West Wycombe Estate with our Bomford Hawk hedge cutter.

Kensham Farms Foreman Nigel Rogers, succeeded by his son Paul

Up until last year Nigel had cut many of our hedges each year since 1962 with various less sophisticated hedge cutters, leaving other hedges to be cut, at significant expense, by a contractor.

While mentioning Nigel Rogers this is an appropriate time for me and my family to thank him for his work and great skill, including his resourceful mechanical ability, for the 58 years since he first worked with me at Kensham Farm - I have often felt that we were both working for the farm, with production of food as our main aim, and that much of our success in all the earlier years had been due to Nigel's work.

In the Clarion of Spring 2014 our editor, Katy Dunn, had been kind enough to put the heading to my quarterly 'On the Land' report as 'Farm Foreman retires after 52 years', at the time when Nigel's son Paul took over as farms foreman responsible for all the arable cropping, working with my son Charlie. However, that did not mean that Nigel had finished working on our farms, it turns out that announcement had been quite premature. One of Nigel's main interests now is to restore tractors from long ago, to bring them back to life and in a fit state to take part in the annual Chilterns Vintage Tractor Run which raises funds for Thames Valley Air Ambulance.

Airfield for High Wycombe and District Model Aircraft Club

It was with one of those restored tractors that Nigel ploughed, cultivated, and prepared the seedbed for our new Airfield for High Wycombe and District Model Aircraft Club (HWDMAC), a well-established club with roots going back to the 1940s which had been displaced in Summer 2019 from its earlier flying site on land that was to be used for other purposes.

Planning issues take a long time for a project of this sort and include levels of noise (found after tests with one of the Cub member's model aeroplane to be far less than the noise from the M40 Motorway nearby) and suitability of this recreational activity in the Chilterns Area of Outstanding Natural Beauty (AONB) administered by the Chilterns Conservation Board. Other factors that had to be taken into account were proximity to roads, motorways, residential houses, power cables and public footpaths, health & safety issues and access from public highways as well as the Club's requirement for a landing strip of at least 100m.

This needed to be on a fairly level field facing southwest into the prevailing wind with no difficult hedgerow trees to cause a hazard to the model aircraft. However, we and the Club Committee and its advisers surmounted all of these obstacles so that following harvest we felt that consent would probably be granted by the Bucks Council (Wycombe District).

September is always a good month in which to sow grass seeds, so we started preparing the seedbed, and ordered the special grass seeds, a mixture of 75% of Creeping Red Fescue, 20% Chewings Red Fescue and 5% Highland Bent, in September. The photos show Nigel and Paul preparing the spinner mounted on the buggy, and then broadcasting the grass seeds on 16th September, to be followed by a light harrowing so as to ideally cover the seeds with about an inch of nice friable soil. The seeds have taken well as shown in a drone aerial shot of the new runway taken by Alex in late afternoon sunlight on 26th October. The new airfield takes up 1.5 acres (0.6 hectares) of our Chequers Manor Top Plain land, on the left as you drive from Cadmore End towards Stokenchurch.

Nigel and Paul Rogers preparing the spinner to broadcast grass seeds to the new airfield.

The new airfield after establishing of grass seeds prior to use by the High Wycombe and District Model Aircraft Club.

The Clarion – *Spring 2021*

The Agriculture Act 2020. Warburtons National Golden Loaf Award. Re-cladding Gable Ends of Barns. Wild Bird Count.

The Agriculture Act 2020

The new Agriculture Act 2020 has been described in the Farm Business weekly bulletin as being *"the biggest shake-up in domestic farm policy in living memory"*.

Readers of this column in The Clarion, which is now in its fourteenth year, may think that it is sometimes more about politics than about farming. But that is a necessity in the modern world, since politicians have many duties, one of which is to make laws and regulation that will ensure that their constituents have access to a reliable supply of good quality food.

The last time there was a new British Agriculture Act was in 1947, when Britain was recovering from the danger and deprivation of World War II, and food was still rationed. So now, 73 years later, the passing of the new Agriculture Act 2020 is a momentous event which, quite appropriately, was enacted by receiving Royal Assent on Remembrance Day, the 11th day of November 2020.

The new Agriculture Act was necessary following the Brexit vote for Britain's withdrawal from the European Union, since for all the years between 1973 and 2020 British farm policy had to be based on EU legislation passed in Brussels.

The significant difference between the 1947 and the 2020 Agriculture Acts has been the change from protecting the role of British farmers as producers of home grown food, to the new role of British farmers, as seen by Defra, being primarily as caretakers of the environment, with Environmental Land Management schemes paying farms to deliver "public goods" in the form of clean water and air, soil health, care of wildlife and adaptation to climate change.

The Agriculture Act 2020 was formulated over a period of almost three years. It started with a most impractical Consultation Paper in February 2018 entitled "Health and Harmony" at the time when Michael Gove was Minister for Defra. Many of us farmers submitted responses to that consultation in which we pointed out to Defra that although farmers are primarily food producers the production of food had hardly received a mention in the Consultation Paper.

The next stage was for Defra to produce the Agriculture Bill, in which food production was still not treated as the main aim. Any Bill has to go through the process of the First and Second Readings in the House of Commons, followed by the Committee Stage, the Report Stage, and finally the Third Reading, then a similar sequence in the House of Lords. In this case of the Agriculture Bill their Lordships suggested most useful amendments concerning the importance of food. During those stages the NFU, led vigorously by its first ever female President, Minette Batters, pressed hard for the importance of growing food to be recognised.

So by the time the Agriculture Act 2020 became law several significant improvements had been made. These included: -

- The importance of Food Security and the need to encourage food production from British farms has been recognised.

- The Act gives Government powers to intervene if there were to be unexpected market disruption.

- A Trade and Agriculture Commission has been formed to scrutinise future Free Trade Agreements and provide MPs with a report as to how each Agreement with different nations will impact British farms, including whether imports would be of such low quality that they would have been illegal if produced in Britain.

Re-cladding Gable Ends of a Barn

We have recommissioned one of our smaller former grain stores, redundant since building new purpose made stores in 2013 and 2020, to become suitable premises for a local arboriculture firm, Penn Tree Services. The photo shows my son Charlie with grain spear, my grandson Alex who manages our diversified enterprises as well as land agency and IT matters, foreman Paul Rogers in the Kramer forklift with man-cage and Nick Perry who is replacing the traditional Yorkshire Boarding (which always had gaps between the planks to allow ventilation for any livestock in the building) with box profile steel sheeting, galvanised and factory painted olive green colour to blend in well with the other farm buildings.

Re-cladding gable end of workshop prior to use by Penn Tree Services. Nick Perry in the man cage, Paul Rogers in the Kramer fork lift, Charlie and Alex in the fore-ground.

Wild Bird Count

As part of the Central Chilterns Farmer Cluster, we have been provided with wild bird feeders and half a tonne of wild bird seed mix to spread and feed farmland birds during the 'winter food gap'. Alex has been busy scattering several buckets of seed mix each week in two different cover strips. During an hour-long twitching session in early January, Nick Marriner of the Chilterns Conservation Board was able to count and identify 117 birds of 10 species including 60 Chaffinches, 2 Skylarks, 13 Yellowhammer and a Brambling to be recorded in February's Big Farmland Bird Count.

Alex spreading bird feed provided by the Central Chilterns Farm Cluster.

Warburtons National Golden Loaf Award

So much for the political side of farming. On the practical side we at Kensham Farms were delighted to have won the National Golden Loaf awarded by Warburtons to its best milling wheat grower for *"recognition of excellence integral to our milling wheat programme"* for wheat from the 2018 harvest. This award was made when all deliveries of 2018 harvest wheat had been completed by Summer 2019, and we should have received our 'Golden Loaf' in January 2020 at Kettering at Warburton's Farmer Forum Meeting.

We market all our grain through a farmer owned cooperative named Openfield, which is Britain's only national grain marketing cooperative – this company is owned by the 4,000 farmer-members that form the cooperative, handling over 4 million tonnes of grain each year, around one third of the consumption of all homegrown wheat. Despite Openfield's size, it has retained the efficiency and flexibility of a far smaller marketing cooperative. It is through Openfield that we have a contract for 2,400 tonnes of our milling wheat to be supplied to Warburtons for breadmaking each year.

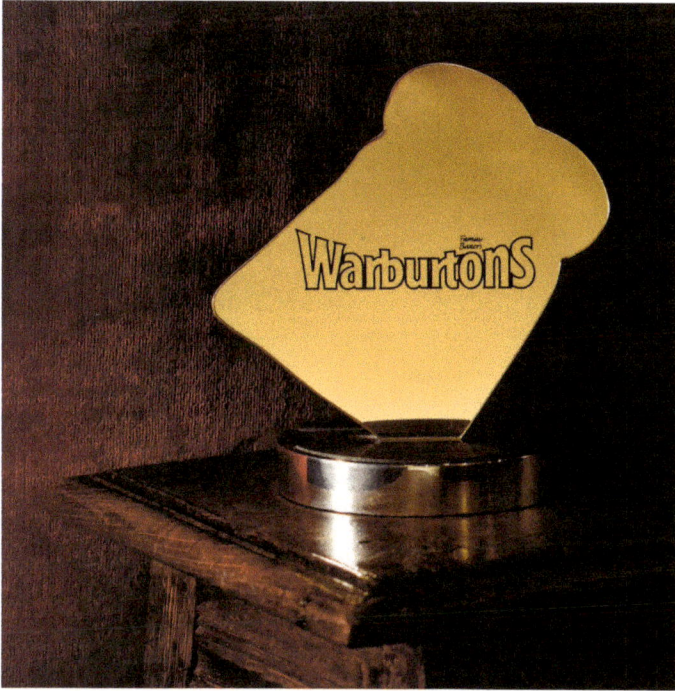

Warburtons Golden Loaf Award for the 2018 crop.

Harvest Festival Hymn,
'We plough the fields – and scatter the good seed on the land but it is fed and watered by God's almighty hand'

Harvest Festivals

Harvest Festival services in Church are normally held towards the end of October; this year in St. Mary-le-Moor Church at Cadmore End there was a packed church on Sunday 17th October, at which the traditional form of Evensong from the 1662 Book of Common Prayer was held. Four of those well-loved Harvest hymns were sung lustily. The best remembered of those hymns was first written in German by Matthias Claudius who lived from 1740 to 1815, more than two centuries ago:-

> **We plough the fields, and scatter**
> **The good seed on the land…..**

One can then ask if anything from two hundred years ago is still relevant in this modern fast-moving age of technology, broadband, manufactured products, and food bought in supermarkets? The answer is that mankind's most basic need for food, and clean water with which to prepare and eat the food, is just the same as it was in those days. And it is a farmer's job to grow it, not only for himself and his family but also for that majority of the population who now live in towns, where there is insufficient land to grow food.

However, over the centuries since the hymn was written farmers have had to tweak the growing process, to provide sufficient food for an enormously increased world population. The main commodity which we now grow at Kensham Farm is 'milling wheat', that is best quality wheat suitable for making bread. It must have high protein and a good level of gluten, so that when it is milled to make flour for bread making the loaf of bread will rise nicely when baked rather than turning out to be stodgy. Some of the detail of growing such wheat nowadays is: -

'We plough the fields….'

In recent years we have tried to reduce use of the plough, since the job of turning all the soil over releases carbon from the soil and is slow and heavy work for the tractor. It consumes a lot of the costly diesel fuel, which emits greenhouse gases that trap heat in the atmosphere and contributes to global warming.

One of the main objectives of ploughing used to be to bury the weeds that had grown up in the preceding crop. Nowadays we try to control any weeds in the crop by spraying with herbicides. If the crop is clean the stubble will be clean without weeds, so that it will no longer be necessary to turn the soil over to bury weeds and trash from the preceding crop. We can use a shallow cultivator on the stubble immediately after harvest to produce a tilth, in which any weed seeds, or 'volunteer' grain which came out of the back of the combine, can chit. Then later in the autumn, late September or October, those young volunteer cereal plants, and any young weed seedlings, can be killed off with Glyphosate spray ('Roundup' is a well know brand) immediately before seeding the new crop.

Defra's Environmental Land Managements Scheme (ELMS), formulated under the Agriculture Act 2020, has three tiers of policy objectives with different scales and scope. The entry level tier that is designed to be accessible to every farm business is called the Sustainable Farming Incentive (SFI). The middle tier that requires groups of farmers working together in a 'Cluster' is called Local Nature Recovery (LNR) and the final tier that will operate over thousands of hectares is called Landscape Recovery.

Dale drill seeding winter wheat into an experimental cover crop of mustard, buckwheat and stubble turnips as part of a Defra SFI pilot scheme. The cover crop has already been killed off with glyphosate.

Kensham Farms is taking part in the SFI pilot before it is launched in 2022 and working with Defra to help shape the scheme and provide feedback on everything from the online application process to soil management practices in the field. To this end we planted around 200 acres of our stubbles with a cover crop of mustard, buckwheat and stubble turnips. This mixture will prevent leaching of nutrients, assist drainage with large tap roots and provide organic matter when incorporated into the soil. Our Tipping Shaw field shown in the photo had already been seeded with this cover crop, which had been grown and was then killed off with Glyphosate spray just before seeding time, all within the space of just over two months after harvest 2021. The photo shows the seed drill at work drilling direct into the cover crop, which has already been killed off, even though it still looks green.

Drone photograph of Tipping Shaw field in which the seeding under the Defra SFI pilot scheme is being carried out.

'… and scatter the good seed on the land'

At the time when the hymn was written the "good seed" meant seed that was as free from weed seeds as was possible in those days. This would indeed have been scattered by hand from a bucket or basket on most farms. But that was to be superseded by a horse-drawn seed drill, the first practical design of which had been perfected in 1701 by Jethro Tull. However, this was not introduced onto most farms until the 1800s.

The design of seed drills has been constantly improving over the years. We now use a Dale seed drill which has a working width of 10 metres, and we use seed which has been cleaned to be free of any contamination prior to being dressed with seed dressing to protect against fungal diseases. The second photo of this was taken in West Wycombe Park on 22nd October, drilling our final field of the 1,650 acres of winter wheat to be seeded this Autumn, ready for harvest in August 2022. We prepared this field without ploughing or cover crop after harvest, just with cultivation of the stubble and Glyphosate spray to kill off any weeds or volunteers.

Drilling winter wheat at West Wycombe Park.

The small octangular building that can be seen in the photo is 'The Temple of the Four Winds', which is a 3-storey tower at the southeast corner of the garden of West Wycombe Park. Our seed trailer and Kramer forklift for loading the one tonne bags of seed into the drill can be seen in front of it.

Nowadays after seeding we care for the health of the crop while it is growing in a way that would not have been possible in earlier times. Such treatments are with fertiliser, lime to neutralise acidity in the soil, spray treatments for control of weeds and fungal diseases, growth regulator to make the stem short and the roots deeper and insecticide against aphids.

Despite all these modern crop protection treatments, the hymn goes on to remind us all those things which have never changed: -

> **But it is fed and watered**
> **By God's almighty hand;**
> **He sends the snow in winter,**
> **The warmth to swell the grain,**
> **The breezes and the sunshine,**
> **And soft refreshing rain**

ACKNOWLEDGEMENTS

As a final note, I would like to thank all my family and many friends who have encouraged me to write this book. It was my late nephew, Emlyn Coldicott, who was the first to suggest that, if I were to set out my experiences at Kensham Farm in the form of a book with some of my photographs, then it might be of interest to others. Emlyn was just starting at primary school at the time when my Alison and I stayed with my sister, Diana, and her family at their house in Farnham Common. On returning from school each day Emlyn used to inspect my progress in making the five sectional pig huts in his parents' garden in August and September 1955, during the few weeks before Alison and I started managing Kensham Farm at Michaelmas 1955. Sadly, Emlyn died just before the manuscript of this book was ready for publication.

My son Charlie has checked the technical detail of all my articles for The Clarion over the past fourteen years, as well as the final manuscript, and without my son Paul's help with the selection of photographs and the process of publication the book might never have been completed. My grandson, Alex Nelms, has been a constant help, working with me in the Kensham Farm office, and typing much of the original copy.

My great thanks are also due to my sister Diana for sharing her experience of writing and publishing several books covering local history in Hampshire, and for checking my first draft of the historical detail of the earlier strip system of farming, followed by the Enclosure Acts, prior to proofreading the final manuscript.

Jo Donachie is a member of the congregation of our local parish church, Holy Trinity, at Lane End, and was the last Education Officer at the Finnamore Wood Young Offender Institution where she also taught many of the lads to whom I was a tutor. I am most grateful to Jo for proofreading the manuscript for me and making many useful suggestions, particularly concerning the integration of the several separate topics with which I have been involved into a suitable sequence in chapter 4.

Finally, my thanks are due to Katy Dunn, editor of The Clarion, for her kind Foreword to Chapter 5 in which she remembered my dear Alison so well, and to Steve Baker for his most generous Foreword to the book, as well as all his work in Parliament over the years as our constituency MP, and his support for farming in the Wycombe area ever since his selection as Conservative candidate for Wycombe in 2009.

January 2022

Bryan Edgley
Kensham Farm